AF772127

94

Advances in Biochemical Engineering/Biotechnology

Series Editor: T. Scheper

Advances in Biochemical Engineering/Biotechnology

Series Editor: T. Scheper

Recently Published and Forthcoming Volumes

Microscopic Techniques
Volume Editor: Rietdorf, J.
Vol. 95, 2005

Regenerative Medicine II
Clinical and Preclinical Applications
Volume Editor: Yannas, I.V.
Vol. 94, 2005

Regenerative Medicine I
Theories, Models and Methods
Volume Editor: Yannas, I.V.
Vol. 93, 2005

Technology Transfer in Biotechnology
Volume Editor: Kragl, U.
Vol. 92, 2005

Recent Progress of Biochemical and Biomedical Engineering in Japan II
Volume Editor: Kobayashi, T.
Vol. 91, 2004

Recent Progress of Biochemical and Biomedical Engineering in Japan I
Volume Editor: Kobayashi, T.
Vol. 90, 2004

Physiological Stress Responses in Bioprocesses
Volume Editor: Enfors, S.-O.
Vol. 89, 2004

Molecular Biotechnology of Fungal β-Lactam Antibiotics and Related Peptide Synthetases
Volume Editor: Brakhage, A.
Vol. 88, 2004

Biomanufacturing
Volume Editor: Zhong, J.-J.
Vol. 87, 2004

New Trends and Developments in Biochemical Engineering
Vol. 86, 2004

Biotechnology in India II
Volume Editors: Ghose, T.K., Ghosh, P.
Vol. 85, 2003

Biotechnology in India I
Volume Editors: Ghose, T.K., Ghosh, P.
Vol. 84, 2003

Proteomics of Microorganisms
Volume Editors: Hecker, M., Müllner, S.
Vol. 83, 2003

Biomethanation II
Volume Editor: Ahring, B.K.
Vol. 82, 2003

Biomethanation I
Volume Editor: Ahring, B.K.
Vol. 81, 2003

Process Integration in Biochemical Engineering
Volume Editors: von Stockar, U., van der Wielen, L.A.M.
Vol. 80, 2003

Microbial Production of L-Amino Acids
Volume Editors: Faurie, R., Thommel J.
Vol. 79, 2003

Phytoremediation
Volume Editor: Tsao, D.T.
Vol. 78, 2003

Chip Technology
Volume Editor: Hoheisel, J.
Vol. 77, 2002

Modern Advances in Chromatography
Volume Editor: Freitag, R.
Vol. 76, 2002

History and Trends in Bioprocessing and Biotransformation
Vol. 75, 2002

Regenerative Medicine II
Clinical and Preclinical Applications

Volume Editor: Ioannis V. Yannas

With contributions by
A. Atala · C. E. Butler · R. M. Capito · J. Fawcett · M. P. Hatton
B. Kinner · J. E. Mayer Jr. · A. G. Mikos · A. S. Mistry · D. P. Orgill
E. Rabkin-Aikawa · P. A. D. Rubin · F. J. Schoen · M. Spector
P. Verma · I. V. Yannas · M. Zhang

Springer

Advances in Biochemical Engineering/Biotechnology reviews actual trends in modern biotechnology. Its aim is to cover all aspects of this interdisciplinary technology where knowledge, methods and expertise are required for chemistry, biochemistry, micro-biology, genetics, chemical engineering and computer science. Special volumes are dedicated to selected topics which focus on new biotechnological products and new processes for their synthesis and purification. They give the state-of-the-art of a topic in a comprehensive way thus being a valuable source for the next 3–5 years. It also discusses new discoveries and applications.

Special volumes are edited by well known guest editors who invite reputed authors for the review articles in their volumes.

In references *Advances in Biochemical Engineering/Biotechnology* is abbreviated as *Adv Biochem Engin/Biotechnol* as a journal.

Visit the ABE home page at springeronline.com

ISBN 978-3-642-06168-4 e-ISBN 978-3-540-31543-8

ISSN 0724-6145

DOI 10.1007/b14096

Springer is a part of Springer Science+Business Media
springeronline.com

Softcover reprint of the hardcover 1st edition 2005

Cover: KünkelLopka GmbH, Heidelberg; design & production GmbH, Heidelberg

Printed on acid-free paper 02/3141xv – 5 4 3 2 1 0

Advances in Biochemical Engineering/Biotechnology Also Available Electronically

For all customers who have a standing order to Advances in Biochemical Engineering/Biotechnology, we offer the electronic version via SpringerLink free of charge. Please contact your librarian who can receive a password for free access to the full articles by registering at:

springerlink.com

If you do not have a subscription, you can still view the tables of contents of the volumes and the abstract of each article by going to the SpringerLink Homepage, clicking on "Browse by Online Libraries", then "Chemical Sciences", and finally choose Advances in Biochemical Engineering/Biotechnology.

You will find information about the

- Editorial Board
- Aims and Scope
- Instructions for Authors
- Sample Contribution

at springeronline.com using the search function.

Attention all Users
of the "Springer Handbook of Enzymes"

Information on this handbook can be found on the internet at
springeronline.com

A complete list of all enzyme entries either as an alphabetical Name Index or as the EC-Number Index is available at the above mentioned URL. You can download and print them free of charge.

A complete list of all synonyms (more than 25,000 entries) used for the enyzmes is available in print form (ISBN 3-540-41830-X).

Save 15%

We recommend a standing order for the series to ensure you automatically receive all volumes and all supplements and save 15% on the list price.

Preface

How do you grow back a disabled organ? The key is to find a way to induce tissue and organ regeneration. This prospect is clearly different from spontaneous phenomena such as compensatory growth of the adult liver or kidney; nor is it the epimorphic regeneration of limbs that is spontaneously observed with certain amphibians. It is the induced regrowth of an organ at the anatomical site of an adult where the function of the original organ has been lost, either following accidental trauma or elective surgery or, conceivably, after an organ has become dysfunctional due to a chronic insult.

This is a new field of study; yet, this volume provides answers from well-known investigators for each of several adult organs. These are early efforts and the regenerated organs are often quite imperfect. Nevertheless, in some cases, the clinical benefit appears to be highly significant; occasionally, even unique and indispensable.

In spite of its rough and undeveloped contours, the methodology of induced regeneration has already yielded spectacular results over the entire anatomy. There are chapters on induced regeneration of heart valves, peripheral nerves, skin, cartilage, urological organs, the conjunctiva, bone, liver, the spinal cord. Useful background is provided by a chapter on the compensatory hypertrophy of the liver, an unusual and instructive healing process in the adult. In addition, there are also chapters that describe other spontaneous regeneration phenomena such as those observed in the mammalian fetus or limb (epimorphic) regeneration observed in amphibians. Background chapters of this type are useful as relevant underpinnings, or even possibly as "controls", for data from the adult mammal. An experimental tool, the use of stem cells, is treated separately. Two emerging theoretical approaches, the fetal healing reactivation theory and the immunocompetence theory, are also described. Methodology and theories have been selected by the editor on the expectation that each will become increasingly useful in tomorrow's investigations.

Regeneration of adult organs is an unexpected fact. It is the mammalian fetus, not the adult, that regenerates spontaneously following severe injury to an organ. Following the remarkable pioneering efforts of the 1950s and 1960s that led to transplantation of the kidney and the heart, it had become widely accepted that loss of adult organ function could only be treated by heroic efforts. A serendipitous discovery was made in the mid-1970s that changed this thinking. Studies with animals, and later with burn patients who had lost large areas of their skin, showed that the healing processes of deep skin wounds could be controlled by use of a protein scaffold, misnamed "artificial skin" at that time. Unexpectedly and dramatically, but quite confusingly for a while, this scaffold changed the adult healing process from contraction and scar formation to partial regeneration of skin. The new skin was imperfect, lacking hair and sweat glands, but it was made up of a normal epidermis and, most surprisingly, a normal dermis as well as a normal epidermal-dermal junction.

It is now clear that at least partial regeneration of several organs can be obtained by the use of no more than three classes of relatively simple "reactants": cell suspensions, solutions of growth factors, and insoluble matrices. The rules that specify the required reactants for "synthesis" of tissues and organs are emerging rapidly. There is evidence that regeneration of the stroma, perhaps the most challenging problem in organ regeneration, does not require use of exogenous cells or exogenous growth factors but only requires grafting at the injured site of a degradable scaffold, a "regeneration template", with a highly specific structure. In general, the methodology of organ regeneration is far simpler than that used for organ transplantation and can be largely defined, even occasionally standardized, using physicochemical and biochemical criteria. The new methodology is often referred to as in vivo tissue engineering. One successful paradigm has consisted of *de novo* organ synthesis at the correct anatomical site using cell-seeded scaffolds. In other cases, organ synthesis has been achieved *in vitro* prior to implantation at the correct anatomical site.

A fascinating possibility emerges after the data on induced organ regeneration have been examined: The fetal response to injury theoretically lies fast asleep in the adult but it can be awakened by use of appropriate tools and harnessed to grow back tissues and organs. Another interesting prospect emerges following the discovery of associations between the fetal processes of loss of regenerative potential and the concurrent acquisition of immunocompetence. These recent theoretical advances provide formidable and exciting challenges to future investigators of organ regeneration in adults.

The editor would like to thank the staff at Springer-Verlag who have helped him during the preparation of this volume. My thanks go to Professor Dr. Thomas Scheper, Institut für Technische Chemie, Hannover University, Germany as well as to Dr. Marion Hertel, Chemistry Editor, and especially to Ms. Ulrike Kreusel, Chemistry Desk Editor.

Cambridge, October 2004 Ioannis V. Yannas

Contents

Contents of Volume 93

Regenerative Medicine I

Theories, Models and Methods

Volume Editor: Ioannis V. Yannas

Adv Biochem Engin/Biotechnol (2005) 94: 1–22
DOI 10.1007/b99997

Tissue Engineering Strategies for Bone Regeneration

Amit S. Mistry · Antonios G. Mikos (✉)

Department of Bioengineering, Rice University, 6100 Main, MS-142, PO Box 1892, Houston, TX 77005–1892, USA
amistry@rice.edu, mikos@rice.edu

Abstract Bone loss due to trauma or disease is an increasingly serious health problem. Current clinical treatments for critical-sized defects are problematic and often yield poor healing due to the complicated anatomy and physiology of bone tissue, as well as the limitations of medical technology. Bone tissue engineering offers a promising alternative strategy of healing severe bone injuries by utilizing the body's natural biological response to tissue damage in conjunction with engineering principles. Osteogenic cells, growth factors, and biomaterial scaffolds form the foundation of the many bone tissue engineering strategies employed to achieve repair and restoration of damaged tissue. An ideal biomaterial scaffold will provide mechanical support to an injured site and also deliver growth factors and cells into a defect to encourage tissue growth. Additionally, this biomaterial should degrade in a controlled manner without causing a significant inflammatory response. The following chapter highlights multiple strategies and the most recent advances in various areas of research for bone tissue regeneration.

Keywords Bone · Mesenchymal stem cell · Growth factor · Biodegradable scaffold · Bioreactor

Abbreviations

MSC	Mesenchymal stem cell
HA	Hydroxyapatite
ECM	Extracellular matrix
OPF	Oligo(poly(ethylene glycol) fumarate)
PGA	Poly(glycolic acid)
VEGF	Vascular endothelial growth factor
BMP	Bone morphogenic protein
TGF	Transforming growth factor
FGF	Fibroblast growth factor
IGF	Insulin-like growth factor
PDGF	Platelet-derived growth factor
EGF	Epidermal growth factor
rh-BMP-2	Recombinant human BMP-2
PDLLA	Poly(D,L-lactic acid)
PLGA	Poly(D, L-lactic-*co*-glycolic acid)
PEG	Poly(ethylene glycol)
PPF	Poly(propylene fumarate)
PPF-DA	Poly(propylene fumarate)-diacrylate
hBMP-2	Human BMP-2
β-TCP	β-Tricalcium phosphate
PLLA	Poly(L-lactic acid)
RGD	Arg-Gly-Asp

1 Introduction

Each year, more than 6.3 million fractures occur in the United States [1], of which almost 1 million require hospitalization [2]. Over 10 million Americans suffer from deteriorating bone as a result of osteoporosis, with another 34 million at risk of developing the disease due to low bone mass [3]. Osteogenesis imperfecta is a genetic disorder affecting up to 50,000 Americans by disrupting either the amount or quality of collagen in bones [4]. Patients with either of these diseases possess weak bones with a high propensity for fracture or severe injury. Bone cancers, primarily osteosarcoma, are treated by tumor resection, which requires restoration of a large volume of bone [5]. In addition, there is a high need for bone repair or replacement in cases of total joint arthroplasty, spinal arthrodesis, maxillofacial surgery, and implant fixation [6, 7].

In all of these cases, bone must be regenerated to fill in a defect and restore structure and function to damaged tissue. Natural healing of bone with mechanical fixation can, in most cases, adequately mend minor fractures over time. However, delayed unions or non-unions have been observed in up to 10% of all fractures [8]. The lack of proper union in these critical-sized defects prevents restoration of function to the damaged bone [9].

Current medical treatments for severe bone injuries are problematic and may yield poor results. The "gold standard" for healing bone defects is transplantation of natural bone tissue from the patient (autograft); however, there are limited sites where bone may be harvested without loss of function [10, 11]. Autografts are less effective in irregularly shaped defects and may be resorbed prior to complete healing. Furthermore, autografts harvested from the iliac crest of the hip are associated with a 10% complication rate including infection, fracture, pain, paresthesia, nerve injury, and donor-site morbidity [12, 13]. Allografts derived from cadavers are another commonly used bone graft material. However, disease transmission and immunologic rejection are serious concerns with unprocessed allografts, and processed allografts, such as demineralized bone matrix, lack bone growth inducing factors necessary for efficacy [14]. Xenografts, or bone grafts obtained from different species, are also a poor option due to the danger of disease transmission or immunological rejection [15].

Metals such as iron, cobalt, and titanium may be permanently placed in bone to fill a defect and provide internal fixation; however, fatigue, corrosion, tissue infection, and poor implant-tissue interface create many problems for patients [16, 17]. Additionally, metals typically possess mechanical properties significantly greater than natural bone, thereby absorbing much of the mechanical stimuli needed for proper regrowth of bone [18]. This stress-shielding effect causes bone resorption around the implant and eventually may require whole-implant removal.

Ceramics are also used in the treatment of bone injuries since their composition is similar to the inorganic composition of bone [19]. While ceramics offer excellent biocompatibility, they are too brittle to provide structural support to load-bearing bones. They also degrade much more slowly than new bone forms, thereby inhibiting regrowth in a defect.

Severe fractures or large-volume bone loss may be treated by distraction osteogenesis, which entails the lengthening of limbs across a defect through temporary external fixation devices [20]. Developing tissue experiences tensile forces from distraction devices as well as compressive forces from activity, resulting in bone growth across a gap. Distraction osteogenesis is a promising strategy for healing critical-sized bone defects especially in conjunction with tissue-engineering principles.

The emerging field of tissue engineering offers therapies for improved healing of damaged tissue without the limitations and drawbacks of current treatments. The goal of tissue engineering is to restore structure and function to a defect by utilizing the body's natural healing response in addition to treatment with one or more of three elements: cells, signaling molecules, and scaffolds [21]. Thus, a bone defect will potentially be replaced by natural bone tissue with complete union and full restoration of function without the use of a permanent implant.

The following chapter describes the challenges involved in achieving this goal from both physiological and engineering perspectives. A variety of tissue-engineering strategies will be presented with specific examples of progress towards regenerative therapies for bone tissue.

2 Anatomy and Physiology of Bone

2.1 Function

The complex organization and incomparable properties of bone tissue enable it to perform a multitude of unique functions in the body. The skeleton is designed to protect vital organs of the body and provide the frame for locomotion of the musculoskeletal system [7]. Both the material properties of bone and the design of whole bones contribute to the exceptional stiffness and strength that give bone the ability to withstand physiological loads without breaking [22]. Additionally, bone is a reservoir for many essential minerals, such as calcium and phosphate, and plays an important role in the regulation of ion concentrations in extracellular fluid [23]. Bone marrow contains mesenchymal stem cells (MSCs), which are pluripotent cells capable of differentiation into bone, cartilage, tendon, muscle, dermis, and fat tissue [24]. Also found in marrow are hematopoietic cells that produce the red and white blood cells that function in nutrient transport and immune resistance, respectively [23].

2.2 Structure and Mechanics

Typically, the adult skeleton contains 80% cortical (compact) bone and 20% trabecular (cancellous) bone [25]. Cortical bone is hard and dense and makes up the shaft surrounding the marrow cavity of long bones as well as the outer shell of some other bones [26]. Cortical bone is only 10% porous, allowing room for only a small number of cells and blood vessels [27]. The structural unit of cortical bone is the cylindrically shaped osteon, which is composed of concentric layers of bone called lamellae [26]. Blood vessels run through Haversian canals located at the center of each osteon while nutrient diffusion is further aided by canaliculi, or microscale canals within bone. Osteons are aligned in the longitudinal direction of bone and therefore, cortical bone is anisotropic. In the longitudinal direction, reported strength values are 79–151 MPa in tension and 131–224 MPa in compression [23]. The moduli range from 17–20 GPa for both tension and compression.

Spongy, porous trabecular bone is found in the ribs, spine, and the ends of long bones [26]. Trabecular bone, which may be as much as 50–90% porous [27], is an interconnected network of small bone trusses (trabeculae) aligned in the direction of loading stress [26]. The porous volume contains vasculature and bone marrow, which provide little mechanical support compared to cortical bone. The strength and moduli of trabecular bone vary with density, but are reported to be between 5–10 MPa and 50–100 MPa respectively (for both tension and compression) [23].

The unique mechanical properties of bone can partly be attributed to the interaction of its chemical components in the nanoscale [28]. Bone is composed of roughly 60% inorganic mineral, 30% organic material, and 10% water [26]. Calcium phosphate crystals, primarily hydroxyapatite (HA), comprise the inorganic phase while the organic phase consists mostly of collagen. Collagen molecules align into triple helices that bundle into fibrils (1.5–3.5 nm diameter), which then bundle into collagen fibers (50–70 nm diameter) [29]. HA crystals are small plates 2–3 nm thick and tens of nanometers in length and width that precipitate onto the collagen fibers [30, 31]. Rigid HA crystals provide compressive strength to the composite, while collagen fibers, capable of energy dissipation, impart tensile properties to bone [32].

2.3 Cellular Organization

Three types of cells inhabit the inorganic–organic composite structure of bone. Osteoblasts, derived from MSCs, secrete collagenous proteins that form the organic matrix of bone, called osteoid [33]. Mature osteoblasts surrounded by osteoid stop secreting this matrix and become osteocytes, which remain important in signal transduction of mechanical stimuli [7, 27]. The third cell type, osteoclasts, are derived from hematopoietic cells of the marrow and secrete acids and proteolytic enzymes which dissolve mineral salts and digest the organic matrix of bone [25, 33]. Osteoblasts and osteoclasts turn over frequently, but operate in a balanced manner such that bone tissue is constantly remodeled in response to various chemical, biological, and mechanical factors [7].

Five phases describe the bone remodeling process: quiescence, activation, resorption, reversal, and formation [34]. The process begins with the quiescence state in which inactive cells are present on the surface of bone. Normally, more than 80% of free bone surfaces are in this state. The activation phase arises as biochemical or physical signals attract mononuclear monocytes and macrophages to the remodeling site and promote differentiation into osteoclasts. Resorption ensues as osteoclasts break down the organic and inorganic components of bone to form a cavity. During reversal, osteoclasts leave the site and mononuclear macrophage-like cells secrete a cement-like substance on the surface. In the final phase, formation, osteoblasts fill the cavity with bone matrix and form new osteons. This begins with the rapid deposition of collagen in densely packed columns, followed by mineralization. As these processes complete, the surface returns to its quiescent state.

2.4 Mechanical Environment

Cell–cell communication is very important for bone maintenance functions. The extracellular matrix (ECM) of bone is a complex milieu of various molecules including peptides, glycoproteins, soluble growth factors, hormones, collagen, and other proteins [27]. Osteocytes lying within the bone matrix play an important role in detecting and then converting mechanical stimuli into biochemical molecules, which signal production or resorption. Bone tissue experiences a variety of mechanical stimuli, including shear forces associated with fluid flow and mechanical loading of whole bones; thus, cellular signals and response are essential for proper maintenance of bone.

2.5 Bone Formation

Three forms of bone formation will be described here: endochondral bone formation, intramembranous ossification, and appositional formation. Endochondral ossification describes the formation of long and short bones during embryonic development, as well as bone formation into a fracture or implanted bone graft [25, 27]. This process begins as MSC progenitors differentiate into chondrocytes, or cartilage-forming cells. These cells form a cartilaginous matrix as they mature, eventually losing the ability to proliferate. Mature chondrocytes then produce calcification proteins while phagocytic cells begin resorption of the cartilaginous matrix. MSCs originating in the periosteum, a tissue layer surrounding long bones, migrate, differentiate, and proliferate in the matrix. These cells are prone to bone (rather than cartilage) formation and subsequently construct passageways for vascularization while continuing to build bone. Osteocytes and randomly oriented collagen fibrils form the bulk of this weak material known as immature woven bone. Remodeling of woven bone into mature, lamellar bone is a slow process that yields a more organized tissue of distinct mechanical strength.

Intramembranous ossification, which applies to the formation of flat bones such as those in the skull, follows a similar process except that bone forms directly from mesenchymal tissue without an intermediary cartilaginous network [23]. Appositional growth occurs as layers of osteoblasts secrete sheets of matrix onto existing bone, resulting in overall bone growth. All three described methods of bone growth occur constantly in the development, growth, and maintenance of bones throughout the body [25].

Another bone maintenance process converts old, concentric lamellar bone into new Haversian systems, or secondary osteons [25]. A sheath of calcified mucopolysaccharides called the cement line divides new and old bone. Osteons outside this line lose connection with the blood supply, and cells within it eventually die. Though the orientation of osteons provides a great deal of strength to bone in compression and tension, if bone does break, it does so along these cement lines rather than across osteons. Thus, cracks in one osteon are not easily propagated through the entire bone.

2.6 Fracture Healing

In addition to a discontinuity within bone, a fracture causes loss of function and damage to blood vessels. Injury initiates a cascade of healing events that recapitulate some of the steps of embryonic bone formation and are described here in three biological stages: inflammation, repair, and remodeling [7, 23]. The initial acute inflammatory response entails the formation of a hematoma at the site of damaged blood vessels. Neutrophils and macrophages arrive and ingest the cellular debris of necrosis while releasing growth factors and cytokines. These biochemical signals induce the migration and differentiation of MSCs from surrounding bone, marrow, and periosteum. Capillary growth and fibroblast activity create fibrovascular granulation tissue at the injury site. As progenitor cells accumulate and differentiate into osteoblasts, a repair blastema is formed. The relatively short inflammatory phase is followed by the repair phase, which begins as osteoblasts rapidly lay down new osteoid to form woven bone at the injury site, now called the bony callus. Gradually, the third phase of healing, remodeling, reshapes and reorganizes collagen fibers forming mechanically strong lamellar bone. Though the remodeling phase may take up to a full year for severe fractures, it returns bone to its original, pre-fracture strength [23]. However, if the body's response to bone injury does not provide enough active cells, or if a defect is too large for the body's natural healing response, non-union will occur at the fracture site [7].

3 Bone Tissue-Engineering Strategies

Currently, there is a high risk and occurrence of fractures and disease-related bone loss in the United States. Severe fractures resulting in non-union of injured bone and bone resection associated with tumor removal cannot be healed by the body's natural healing response, and therefore require medical intervention. However, the most advanced treatments available are limited in effectiveness and often result in complications. Thus, there is a significant need for an alternative strategy for the treatment of severe bone loss or fracture. Ideally, treatment will mimic or enhance the body's natural response to bone injury through the use of cells and bioactive growth factors. Additionally, this remedy will result in natural, mechanically sound bone at the site of injury, eliminating the need for a permanent material that may impede the natural growth of bone.

The emerging field of tissue engineering aims to combine engineering technology and the principles of biological science to develop strategies for the repair and regeneration of lost or damaged tissue. Tissue engineering strategies fall into three general categories: (1) cell-based strategies; (2) growth-factor-based strategies; and (3) matrix-based strategies [7]. In practice, however, most experimental work implements two or more of these strategies together towards a solution. For example, a biodegradable scaffold system may be designed to carry cells into a defect while releasing growth factors into the surrounding tissue as it degrades. In the realm of bone tissue engineering, these strategies require interaction between osteogenic, osteoinductive, and osteoconductive elements. Osteogenic components include cells capable of bone production such as osteoprogenitor cells or differentiated osteoblasts. Osteoinductive factors include bioactive chemicals that induce recruitment, differentiation, and proliferation of the proper cell types at an injury. A material that supports bone growth on itself demonstrates osteoconductivity. An osteoconductive scaffold may provide mechanical support, sites for cell attachment and vascular ingrowth, and a delivery vehicle for implanted growth factors and cells [7].

For successful regeneration of bone tissue at a defect, a potential tissue-engineering solution must fulfill many design requirements. These requirements are outlined here [7, 19, 35]. A potential bone tissue-engineering device must:

1. Provide temporary mechanical strength to the affected area.
2. Act as a substrate for osteoid deposition and growth.
3. Contain a porous architecture to allow for vascularization and bone ingrowth.
4. Encourage bone cell migration into a defect and enhance cell activity for regeneration and repair.
5. Degrade in a controlled manner to facilitate load transfer to developing bone and to allow bone growth into the defect.
6. Degrade into non-toxic products that can safely be removed by the body.
7. Not cause a significant inflammatory response.
8. Be capable of sterilization without loss of bioactivity.

The following sections of this chapter describe current, promising strategies employed in bone tissue-engineering research and how they address these design requirements.

3.1 Cell Transplantation

Cell-based strategies for bone tissue engineering involve the transplantation of osteogenic cells into a defect. This form of treatment may aid in the healing of fractures, but may also be useful for treatment of diabetic, osteoporotic, or aged patients whose cells are less bioactive [7]. Cell transplantation is particularly beneficial to patients with a low number of cells, as in the case of vascular disease or irradiated tissue around the site of tumor resection. Cells used for cell-based therapies may be transplanted in various forms including fresh bone marrow, MSCs expanded in culture, or osteoblasts differentiated in culture.

Bone marrow may be isolated from the iliac crest of a fracture patient and then injected into an injury without morbidity at the donor site. Werntz et al. demonstrated the osteogenic capability of autologous bone marrow in segmental bone defects of rats [36]. Live marrow was implanted into femoral defects and resulted in a rate of union comparable to autologous bone graft. Clinically, Connolly et al. harvested marrow from the iliac crest of patients and injected it into unconnected tibial fractures with fixation by cast or nails [37]. Bone union occurred in 18 out of 20 fractures, demonstrating efficacy equal to that of autografts, without the complications typically associated with them.

The key factor for the success of bone marrow in healing non-unions is the presence of MSCs. These progenitor cells are promising tools for regenerative therapy due to proliferation capabilities, preservation of bioactivity after freezing, and the ability to build new tissue in a defect [24]. However, these cells make up less than 0.001% of the cellular content of bone marrow, and even less as age increases [38]. The limited quantity of MSCs in marrow has led to the development of methods to isolate progenitor cells from bone marrow and expand them in vitro. Bruder et al. demonstrated that isolated MSCs can undergo over 30 passages, or more than a billion-fold expansion, without losing osteogenic potential [39]. Hence, these cells may be isolated from a patient, expanded in culture, and seeded onto a carrier for implantation into a defect (Fig. 1).

Richards et al. demonstrated the osteogenic capabilities of cultured MSCs in a collagen gel carrier injected into distraction gaps of rats femora [40]. Significantly more bone formed in gaps with MSC injections than gaps with cell-free injections. Kadiyala et al. loaded MSCs onto ceramic cylinders and implanted them into critical-sized defects in the femora of rats [41]. Cell/ceramic constructs achieved bone formation after 4 weeks and complete union through the

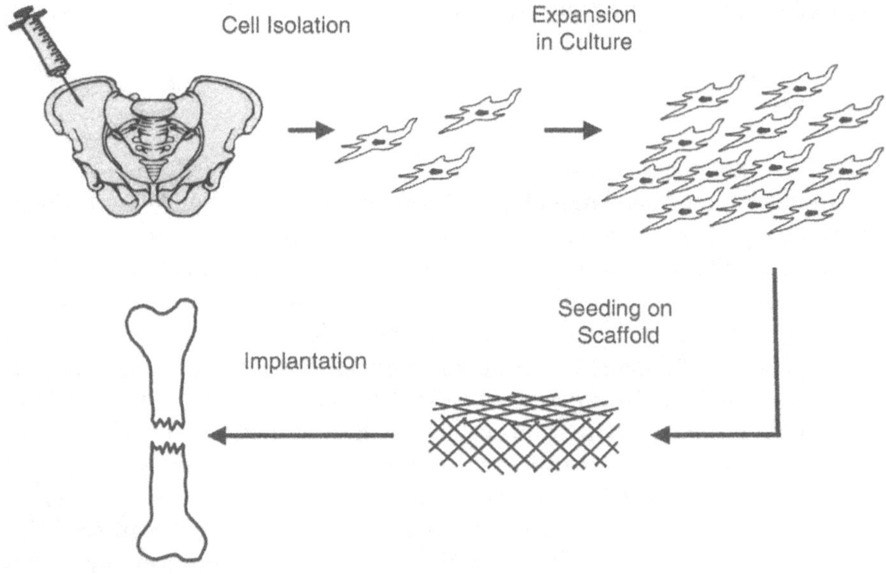

Fig. 1 Cell transplantation: osteogenic cells may be isolated from the iliac crest, expanded in cell culture, seeded onto a scaffold, and implanted into a defect

implant in 8 weeks, while control implants with and without whole marrow did not achieve union. Bruder et al. observed similar results in a canine model segmental femoral defect [42]. The major drawback to these experiments, however, is the slow degradation of ceramic implants in bone.

As mentioned previously, pluripotent MSCs are capable of differentiation into cells of a variety of musculoskeletal tissues. Isolated MSCs in culture can be selectively differentiated into osteoblasts with media supplements such as dexamethasone, ascorbic acid, and β-glycerophosphate [39, 43]. Therefor, MSCs can be harvested from the marrow, expanded in culture, and differentiated into osteoblasts prior to implantation. These osteogenic cells can begin building new osteoid immediately upon arrival at an injury site. Biomaterial carriers, such as biodegradable polymers and ceramics, have been used to deliver differentiated MSCs to a defect site. Temenoff et al. used a novel injectable and biodegradable polymer, oligo[poly(ethylene glycol) fumarate] (OPF), as a cell carrier [44]. Rat marrow stromal cells were encapsulated within the hydrogel and cultured with osteogenic supplements. MSC differentiation was confirmed by matrix mineralization in the crosslinked hydrogels. Yoshikawa et al. observed rapid bone formation by rat marrow stromal cells cultured on porous HA scaffolds with osteogenic supplements [45]. Upon subcutaneous implantation in rats, biochemical and histological analysis showed bone formation occurring at a faster rate in these implants than in those with either fresh bone marrow or undifferentiated MSCs. In another study, Breitbart et al. demonstrated that progenitor cells from the periosteum cultured with dexamethasone could be delivered to a rabbit cranial defect via a degradable poly(glycolic acid) (PGA) scaffold. These cells generated a significant quantity of bone compared to cell-free implants or untreated defects [46].

3.2 Bioreactors

In all of the experiments described in the previous section, cells were cultured on scaffolds by traditional monolayer culture in static flasks, which limits three-dimensional cell interactions and activity. This cell culture method relies solely on molecular diffusion for transport of nutrients, waste, and biochemical signals between cells and media, and may not satisfy the metabolic needs of cells seeded deep within a scaffold for an extended period of time [47]. Bioreactors that simulate the dynamic environment that cells encounter in vivo can improve mass transfer throughout a scaffold to address these concerns [48]. These devices uniformly distribute cells onto three-dimensional scaffolds with appropriate nutrient concentrations, facilitate mass transfer to growing cells, and impart mechanical stimuli to these cells [49]. Bancroft et al. designed a flow perfusion bioreactor that allows culture media to perfuse through scaffolds based on the principle that flow diminishes nutrient concentration gradients and improves medium transport through porous scaffolds while also applying shear stress to developing cells (Fig. 2) [50]. Working with marrow stromal osteoblasts seeded onto titanium fiber meshes, dramatic increases in calcified matrix production and total calcium content were observed compared to statically cultured controls. Further studies by Bancroft et al. demonstrated that flow perfusion increased mineralized matrix production, matrix remodeling, and cell/matrix distribution [51]. In similar studies with murine K8 osteosarcoma cells cultured in collagen sponges in a flow perfusion bioreactor, Mueller et al. demonstrated improved alkaline phosphate activity, mineralization, and overall cell viability in comparison to static culture [52]. Sikavitsas et al. investigated the specific effect of fluid shear stress with constant transport of nutrients in a flow perfusion bioreactor and found that increases in fluid shear stress resulted in enhanced mineralization, which indicates MSC differentiation along the osteoblast lineage [53]. Thus, perfusion bioreactors enhance nutrient transport and, perhaps more importantly, impart shear forces onto cells that encourage differentiation and cell activity.

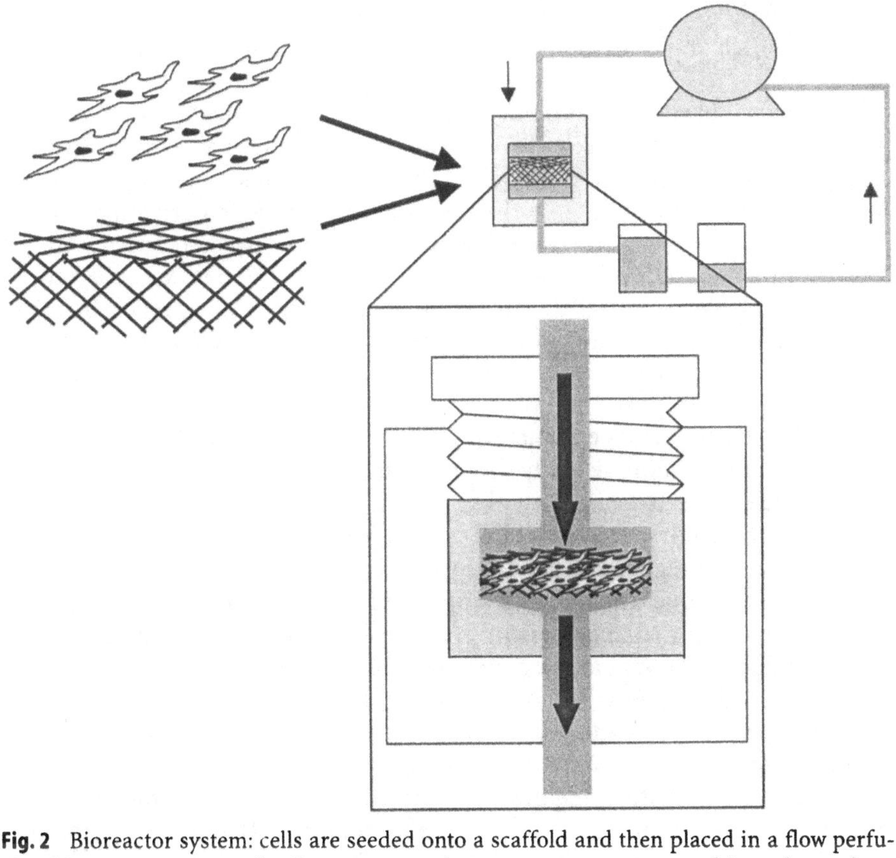

Fig. 2 Bioreactor system: cells are seeded onto a scaffold and then placed in a flow perfusion bioreactor system. The flow system enhances mass transport and introduces shear stress onto developing cells to induce differentiation, proliferation, and mineral deposition

3.3 Growth Factor Delivery

Growth factors are signaling polypeptide molecules that regulate a multitude of cellular functions including proliferation, differentiation, migration, adhesion, and gene expression [54]. Growth factors bind to specific surface receptors of target cells to induce a response, usually in the form of new mRNA or protein synthesis. These biomolecules exhibit pleiotropy, that is, a single growth factor may induce the same or different response in various cell types. In addition, many cell types secrete the same growth factors, though not always with the same effect. Another property of growth factors is redundancy, whereby different growth factors produce the same biological response. In some cases, growth factors produce a more significant response at higher concentrations [55] or produce completely different responses at high and low concentrations [25]. Growth factors also have the ability to enhance or impede the production or activity of other growth factors [54]. The major challenge of working with growth factors is a relatively short biological half-life, which can be as low as 2 min [56]. For applications in humans, high doses of growth factors are required, which are unfortunately associated with high costs and limited supply [7].

A great deal of research is being conducted on growth factors and delivery mechanisms that sustain release, and therefore maintain activity of growth factors. For bone applications, much of this work has been focused on osteoinductive factors; however, growth factors promoting vascularization are arguably of equal importance. Growth factors have been isolated and identified that promote vasculogenesis, angiogenesis, or both [57, 58]. Vasculogenesis refers to embryonic vessel development from endothelial precursors that differentiate and expand into blood vessels, smooth muscle cells, pericytes, and fibroblasts. Angiogenesis is the remodeling and maturation of these vessels that contributes to post-embryonic blood vessel growth and maintenance. Originally, growth factors enhancing angiogenesis were associated with tumor metastasis, though more recently they have found practical use in regenerative therapies [59]. For example, vascular endothelial growth factor (VEGF) is a cytokine that encourages endothelial cell proliferation and migration, induces vasodilation, and regulates both angiogenesis and vasculogenesis [60, 61]. In fact, VEGF is unique among most angiogenic growth factors in that it acts directly upon endothelial cells. Peng et al. observed that VEGF combined with an osteoinductive growth factor enhanced bone formation in cranial defects of rats compared to either growth factor alone [62].

Though many osteoinductive growth factors have been identified, two of the most common groups are the bone morphogenic proteins (e.g., BMP-2, BMP-7) and the transforming growth factor-βs (i.e., TGF-β1), which are both part of the transforming growth factor-β superfamily [54]. BMPs are cytokines that stimulate MSC differentiation into osteoblasts as well as proliferation and function of both chondrocytes and osteoblasts [25, 27]. TGF-βs are growth factors that stimulate MSC differentiation towards chondrocytes and proliferation of osteoblasts and chondrocytes. Ironically, they have also been shown to enhance bone resorption at certain concentrations [25, 27]. Other growth factors with similar effects are: fibroblast growth factors (FGFs), insulin-like growth factors (IGFs), platelet-derived growth factors (PDGFs), and epidermal growth factors (EGFs) [27].

The potent bioactivity of BMPs was first applied by Marshall Urist in 1965 to induce ectopic bone formation in muscle pouches of rabbits, rats, mice, and guinea pigs [63]. Since then, various BMPs have been isolated, characterized, and cloned [64]. After numerous animal studies and human clinical trials demonstrating osteoinductive properties on a par with or better than bone autograft, the Food and Drug Administration (FDA) recently approved recombinant human BMP-2 (rhBMP-2) for use in spinal fusion procedures. Other applications where BMPs may be useful are osteochondral defects, implant stabilization, and the treatment of osteoporosis [65].

As previously mentioned, growth factors have short half-lives and require high dosages for effectiveness. Thus, a delivery device should be chosen to sustain release of the molecules and achieve maximum activity [54]. The carrier should also localize growth factor release to the injury site by direct implantation in or around the affected tissue. Injectable or moldable biomaterials offer a great advantage in this regard as they can easily be fitted into irregularly shaped defects. These carrier materials should be biocompatible, easy to handle and sterilize, and degradable so as not to interfere with bone growth at the defect site [66].

Biomaterials chosen as carriers may be natural or synthetic, and degradable or non-degradable. Natural polymers, such as collagen, can be extracted from animals, treated to block immunogenicity, and used in human applications. Collagen has shown promise as a carrier for BMP, as it is a natural material accepted by the body and is capable of sustained release of growth factors [66]. Biodegradable polymers are also widely studied as delivery vehicles for growth factors for many reasons [54]. Protein structure is maintained in polymeric devices, thus maintaining bioactivity. Delivery rate and duration can easily be controlled through growth factor loading, polymer properties, and processing conditions. Growth factors in polymeric systems can either be incorporated into the biomaterial during processing or loaded into the material after fabrication. If incorporated into a material, growth factors will be released in a diffusion-controlled manner as the material degrades. Alternatively, growth factors may be encapsulated in microparticles, nanoparticles, fibers, or other materials and then incorporated into a matrix for delivery. Another option is the coating of biomaterials with growth factors to facilitate integration into bone [67].

Poly(α-hydroxy esters) have been extensively studied as carriers for osteoinductive growth factors in the form of scaffolds as well as microparticles. Wheeler et al. used poly(D,L -lactic acid) (PDLLA) scaffolds loaded with rhBMP-2 to treat critical-sized defects in rabbit radii [68]. Results showed that rhBMP-2-treated bones performed comparably to autograft-treated bones in biomechanical tests compared to untreated controls. Holland et al. loaded TGF- β1 in gelatin microparticles that were subsequently encapsulated in OPF hydrogels to control release of the growth factor [69]. Observed release profiles showed considerable control over the duration and dosage of TGF-β1 delivery through adjustments in polymer formulation and loading quantities. Though this system was designed for treatment of articular cartilage, the ability of this mechanism to achieve sustained, low-dose delivery of growth factors is applicable to bone and many other tissues.

Hedberg et al. demonstrated loading of TP508, a synthetic peptide capable of influencing bone regeneration, in poly(D,L-lactic-*co*-glycolic acid)/poly(ethylene glycol) (PLGA/PEG) blend microparticles [70]. TP508-loaded microparticles were then loaded into porous, injectable, in situ crosslinkable poly(propylene fumarate)/poly(propylene fumarate)-diacrylate (PPF/PPF-DA) polymer for delivery. Modifications in carrier composition and TP508 loading quantity resulted in variations in peptide release from 53±4% to 86±2% of the total encapsulated quantity, thus demonstrating the experimenter's control over the release of TP508.

3.4 Gene Therapy

Though the previously described delivery systems are capable of achieving sustained release of growth factors, the high dosage requirements for bone applications are still of great concern. A novel approach to this problem is to genetically modify cells to produce osteoinductive growth factors and seed these cells into a scaffold (Fig. 3) [71]. Upon implantation, this cell/scaffold construct will achieve endogenous production of the desired bioactive molecules at a defect site. Thus, cell transplantation and drug delivery techniques can be used together to facilitate bone regeneration. Cell transfection may be accomplished in vivo by introduction of a vector directly into the tissue of interest, or, ex vivo by genetic modification of cells in culture prior to transplantation [72]. Either mechanism can achieve modification of cells for sustained delivery at the site of injury, though there are safety risks associated with in vivo gene therapy and complicated methods involved in ex vivo gene therapy.

Various mechanisms can be utilized to tranfect cells with desired genes for osteoinductive protein production. Nonviral vectors such as cationic polymers and liposomes are safe options, but have low transfection rates and limited efficacy [71]. Certain safety risks are associated with viral gene therapy, even though viral vectors are modified to prevent uncontrolled replication. Blum et al. investigated adenovirus, retrovirus, and cationic lipid vectors for ex vivo genetic modification of rat marrow stromal cells to produce human BMP-2 (hBMP-2) [73]. Cells were isolated, cultured, and genetically modified prior to in vitro and in vivo evaluations of the three transfection methods. In vitro, only adenovirus-modified cells showed significant production of hBMP-2 and markers for differentiation towards osteoblasts. Adenovirus-modified cells seeded

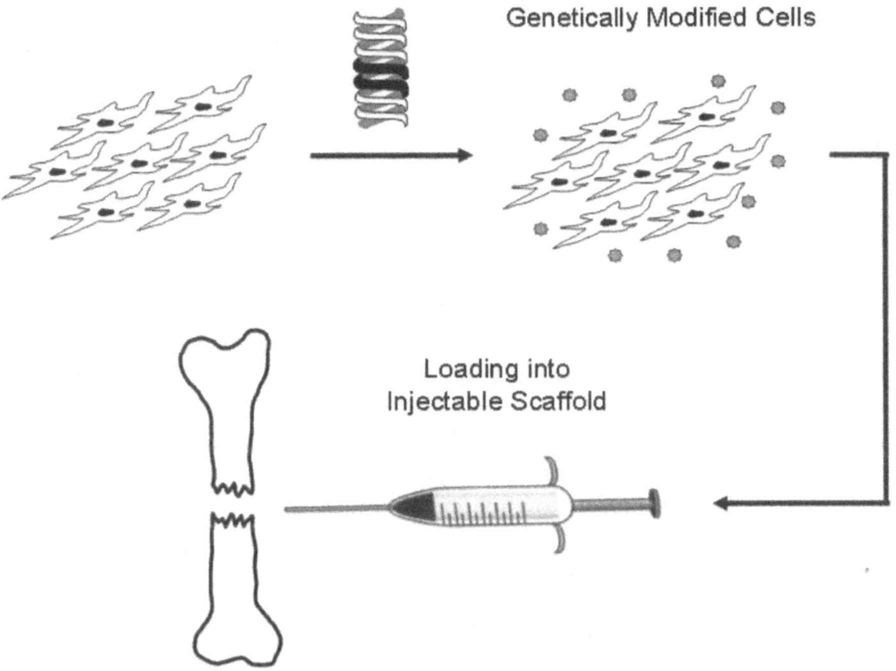

Fig. 3 Gene therapy for growth factor delivery: isolated cells may be genetically modified to produce a desired signalling molecule. Implantation of modified cells loaded into an injectable scaffold can achieve endogenous growth factor delivery at a defect site

onto titanium meshes and implanted into rat cranial defects demonstrated improved bone formation compared to the other two vectors and unmodified cells. Alden et al. further demonstrated the effectiveness of adenovirus with the BMP-2 gene for ectopic bone formation in rats [74]. Viral solution was injected directly into muscle tissue resulting in cartilage formation followed by gradual mineralization and bone formation. Similarly, Baltzer et al. injected adenovirus containing the genetic code for BMP-2 directly into femoral defects in rabbits (with plate fixation) and demonstrated bone healing in 7 weeks compared to non-union in control experiments [75]. Despite these promising results, an immune response against these vectors occurred in all animal experiments representing a significant hurdle for clinical applications of viral gene therapy.

3.5 Scaffolds for Tissue Engineering

Cell-based and growth-factor-based strategies provide the osteogenic and osteoinductive components of a potential treatment for severe bone injury or loss. However, a scaffold biomaterial is essential for filling a critical-sized defect and as a carrier for the cells and/or growth factors used to heal the defect.

In addition, a scaffold must provide temporary mechanical support, osteoconductivity, a porous architecture, controlled degradation capabilities, biocompatibility of itself and of its degradation products, and sterilizability. Each of these factors will be discussed from a design standpoint before discussion of current scaffold-based strategies in practice.

3.5.1 Design Requirements

Mechanical properties of a scaffold should initially match the properties of the target tissue to provide structural stability to an injury site [23]. In this case, the chosen biomaterial must be strong enough to support the physiological load of the body without absorbing the mechanical stimuli required for natural growth in the affected area. It is also essential that the biomaterial allow gradual load transfer to the developing tissue for proper healing to occur [76].

Osteoconductivity refers to a material's ability to act as a substrate for cell adhesion and function while facilitating bone growth throughout a three-dimensional scaffold across a defect. Thus, a porous architecture that allows bone ingrowth and vascularization throughout the scaffold is a necessity. Ideally, scaffolds will be designed to maximize porosity while maintaining mechanical properties with interconnected pores of approximately 200–400 nm [77, 78]. This type of architecture will provide a large surface area for cell attachment, growth, and function, as well as a large void space for bone formation and vascularization [76].

Biodegradable biomaterials are highly preferred over nondegradable materials, even though they are generally weaker. Controlled degradation of a scaffold allows for gradual load transfer to bone, increasing space for bone growth, and eventual filling of a defect with natural bone, as opposed to a permanent biomaterial which may cause stress-shielding or infection [76]. An injectable biomaterial offers the additional advantage of easy fabrication of irregularly shaped implants by molds or by direct injection into a defect site followed by in situ crosslinking (Fig. 3) [79].

The scaffold also must not elicit a significant inflammatory response that may result in fibrous capsule formation around the implant, rendering it ineffective. Toxic response of the tissue surrounding a biomaterial will cause cell death and worsen the injury [76, 80]. The same must hold true for unreacted monomers, initiators, stabilizers, cross-linking agents, emulsifiers, solvents, and degradation by-products released from the scaffold during its implant lifetime.

Finally, the biomaterial scaffold must be sterilizable without loss of function, that is, the sterilization process must not alter the material's chemical composition, as this may affect its bioactivity, biocompatibility, or degradation properties [81].

3.5.2
Biomaterial Selection

A wide variety of polymeric and ceramic materials have been studied for use as biomaterials for bone tissue engineering. Ceramics have been used since 1892 as bone cement [82] as they are biocompatible, osteoinductive, and moldable [83]. The two most heavily studied ceramics are HA [$Ca_{10}(PO_4)_6(OH)_2$] and β-tricalcium phosphate [β-TCP, $Ca_3(PO_4)_2$]. Both materials encourage bone growth, though HA is more osteogenic while β-TCP degrades faster [19]. In a 3-month in vivo study, HA and β-TCP cylinders were implanted in rabbit tibias [84]. Only 5.4% of HA implants resorbed while 85% of β-TCP implants resorbed in the same time frame. The improved ostegenic ability of HA may be attributed to its similarity in composition to natural bone. On the other hand, β-TCP is a multicrystalline, porous ceramic and therefore behaves differently in degradation [85]. Various formulations of these two ceramics, including a biphasic mixture of the two, have been tested with proven biocompatibility and osteoinduction, but with limited structural integrity [86–88]. In addition, ceramic composites from both HA and β-TCP have been used for effective delivery of TGF-β, fibronectin, and cells to defects [23]. The major drawbacks to ceramic materials, however, are their brittleness and slow degradation.

Polymers offer one major advantage over ceramics: flexibility. The mechanical and degradation properties of a polymer can be modified by composition and processing conditions [19]. Factors such as molecular weight, hydrophobicity, and crystallinity are all within the experimenters' control and can be tailored for specific applications [23]. Some polymers can be polymerized in situ via photo- or thermal crosslinking within irregular defects [89]. Though sterilization is more difficult and biocompatibility may be a more significant issue than it is with ceramics, biodegradable polymers represent a very promising class of biomaterials for bone tissue-engineering applications.

Polymers that have frequently been studied for bone tissue engineering include poly(α-hydroxy esters), such as poly(L-lactic acid) (PLLA), PGA, and PLGA, as they have been approved for human use in certain applications by the FDA. Each of these polymers, as well as various modifications or combinations of them, are capable of delivering cells or growth factors to target tissues while also providing a three-dimensional scaffold for cell function [23]. They degrade by hydrolysis into biocompatible components, but lack the mechanical properties required for load-bearing applications. Poly(propylene fumarate) (PPF) is a linear, unsaturated polyester capable of in situ photo- and thermal crosslinking [90]. PPF degrades via ester hydrolysis into biocompatible products – fumaric acid and propylene glycol – and elicits only a mild inflammatory response in rats and rabbits [90, 91]. The mechanical properties of this polymer are also very promising, though not sufficient for cortical bone applications. These properties and degradation rates can be varied based on composition, synthesis conditions, and crosslinking conditions, making PPF easy to tailor for specific applications [79, 89, 92].

The design of a material that is both mechanically strong enough for bone replacement and biodegradable is an especially difficult challenge for scaffold-based strategies. A potential solution to this problem may lie in the reinforcement of polymers with nanomaterials. Ceramic nanoparticles, carbon nanotubes, HA nanocrystals and nanofibers may be incorporated into polymers to improve the mechanical properties of these polymers. Inorganic nanoparticle fillers have been shown to add tensile strength, stiffness, abrasion resistance, crack resistance, and stability to polymer networks [93]. Nonetheless, a major challenge to this work is integration of hydrophilic nanoparticles into hydrophobic polymers. While this is a new and developing field, some progress has been made. Kikuchi et al. combined HA nanocrystals and xenogenic collagen to form a composite with a self-assembled structure similar to bone [94]. However, their material demonstrated significantly weaker mechanical properties in bending than bone. Horch et al. incorporated an aluminum oxide-based ceramic nanoparticle into PPF/PPF-DA and observed significant increases in flexural modulus [95]. The combination of hydrophilic nanoparticles into a hydrophobic polymer was achieved by the addition of surfactant and reactive moieties onto nanoparticles to facilitate dispersion and increase interaction with PPF polymer chains.

Also on the frontier of current bio-nanotechnology research is the synthesis of long peptide chains. Production of synthetic collagen in the nanoscale will not only eliminate the need for xenogenic collagen, but will also introduce unique and novel methods of composite synthesis and nanoscale interactions. Hartgerink et al. accomplished collagen production through pH-induced self-assembly of peptide-amphiphiles [96]. Their group was able to create the organic phase of bone on the nanoscale and utilize this to direct HA crystal growth, thereby forming a nanocomposite that mimics the lowest level of hierarchical organization of bone.

3.5.3 Biomimetic Materials

The development of biomimetic materials is another frontier of tissue-engineering research. These are materials chemically or physically modified for biomolecular recognition by cells at appropriate target sites. Biomimetic materials can enhance cellular recognition and elicit specific cellular responses to aid in tissue growth. This is accomplished through incorporation of cell-binding peptides onto the surface or in the bulk of biomaterials to attain control over cell-biomaterial interactions [97]. Surface modification enhances initial cell reaction while bulk modification is useful for injectable or surface-degrading materials, or materials that require cell interactions throughout the scaffold volume.

The most common peptide for biomimetic surface modification is Arg-Gly-Asp (RGD), an ubiquitous peptide present on ECM proteins and shown to promote cell adhesion in multiple cell types [98, 99]. These peptides can be chemically attached to polymers to facilitate cellular interactions at an injury site. Specifically, RGD peptides have been shown to enhance differentiation, proliferation, and mineralization when attached to the surface of various biodegradable materials [100–103]. Biomimetic PLA and PLGA scaffolds modified with RGD peptides have been prepared and shown to promote the attachment and growth of osteoblasts [102].

Since RGD peptides interact with multiple cell types, there is a great need to identify peptide sequences that elicit more specific responses from selected cell types [104]. Osteopontin is an ECM protein important in adhesion, remodeling, and osseointegration at the biomaterial/tissue interface that is commonly found surrounding mineralized tissues, thus suggesting a more specific interaction with bone tissue [105]. Shin et al. modified OPF hydrogels with an osteopontin-derived peptide and evaluated osteoblast activity on these hydrogels [98]. Results showed that these hydrogels influenced osteoblast proliferation and migration, demonstrating the high potential of these biomimetic materials for tissue-engineering applications.

4 Concluding Remarks

The bone tissue engineering strategies described here all arose from an increasing need for improved treatment of severe bone loss or injury. Osteogenic cells can be isolated, expanded in culture, and transplanted as osteoprogenitor cells or as differentiated osteoblasts. Bioreactor technologies allow for improved cell distribution, nutrient mass transfer, and mechanical stimulation for cells on a scaffold. Growth factors can be delivered to a defect through polymeric carriers, such as microparticles or hydrogels, or produced endogenously through genetically modified cells. A biodegradable polymer or novel composite material with mechanical strength and controllable degradation may be used to fill a defect and deliver cells or signaling molecules to an injury. This scaffold can be prefabricated, injected into a defect for in situ crosslinking, or modified with cell recognition peptides to further aid its function. The experimental examples described in this chapter represent the latest advances in bone tissue engineering from a variety of approaches. Ultimately, many of these strategies may be used together and in conjunction with the body's natural healing response to achieve bone regeneration at the site of a defect.

Acknowledgments The work on bone tissue engineering has been supported by the National Institutes of Health (R01 AR48756) (AGM). A.S. Mistry also acknowledges financial support by a National Institutes of Health Biotechnology Training Grant (T32-GM08362).

References

1. American Academy of Orthopaedic Surgeons (2003) Facts about fractures. In: American Academy of Orthopaedic Surgeons website http://www.aaos.org/wordhtml/research/stats/factshtm. Cited 18 November 2003
2. Hall MJ, Owings MF (2002) 2000 National hospital discharge survey. http://www.cdc.gov/nchs/data/ad/ad329.pdf. CDC National Center for Health Statistics
3. National Osteoporosis Foundation (2003) Fast facts on osteoporosis. In: National Osteoporosis Foundation website http://www.nof.org/osteoporosis/diseasefacts.htm. Cited 18 November 2003
4. Osteogenesis Imperfecta Foundation (2003) Fast facts on osteogenesis imperfecta. In: Osteogenesis Imperfecta Foundation website http://www.oif.org/site/PageServer?pagename=FastFacts. Cited 18 November 2003
5. American Cancer Society (2003) What are the key statistics for bone cancer? In: American Cancer Society website http://www.cancer.org. Cited 18 November 2003
6. Oldham JB, Lu L, Zhu X, Porter BD, Hefferan TE, Larson DR, Currier BL, Mikos AG, Yaszemski MJ (2000) J Biomech Eng 122:289
7. Bancroft GN, Mikos AG (2001) Bone tissue engineering by cell transplantation. In: Ikada Y, Oshima N (eds) Tissue engineering for therapeutic use 5. Elsevier, New York, p 151
8. Cattermole HR, Hardy JR, Gregg PJ (1996) Br J Sports Med 30:171
9. Holy CE, Fialkov JA, Schoichet MS, Davies JE (2000) In vivo models for bone tissue-engineering constructs. In: Davies JE (ed) Bone engineering. em squared, Toronto, p 496
10. Brown KL, Cruess RL (1982) J Bone Joint Surg Am 64:270
11. Enneking WF, Eady JL, Burchardt H (1980) J Bone Joint Surg Am 62:1039
12. Gitelis S, Saiz P (2002) J Am Coll Surg 194:788
13. Younger EM, Chapman MW (1989) J Orthop Trauma 3:192
14. Bostrom RD, Mikos AG (1997) Tissue engineering of bone. In: Atala A, Mooney DJ (eds) Synthetic biodegradable polymer scaffolds. Birkhäuser, Berlin, Germany, p 215
15. Erbe EM, Marx JG, Clineff TD, Bellincampi LD (2001) Eur Spine J 10 [Suppl 2]:S141
16. Petty W, Spanier S, Shuster JJ, Silverthorne C (1985) J Bone Joint Surg Am 67:1236
17. Petty W, Spanier S, Shuster JJ (1988) J Bone Joint Surg Am 70:536
18. Bobyn JD, Mortimer ES, Glassman AH, Engh CA, Miller JE, Brooks CE (1992) Clin Orthop 274:79
19. Temenoff JS, Mikos AG (2000) Biomaterials 21:2405
20. Richards M, Wineman AS, Alsberg E, Goulet JA, Goldstein SA (1999) J Biomech Eng 121:116
21. Langer R, Vacanti JP (1993) Science 260:920
22. Currey JD (2002) Bones: structure and mechanics. Princeton University Press, Princeton
23. Yaszemski MJ, Payne RG, Hayes WC, Langer R, Mikos AG (1996) Biomaterials 17:175
24. Bruder SP, Caplan AI (2000) Bone regeneration through cellular engineering. In: Lanza RP, Langer R, Vacanti J (eds) Principles of tissue engineering. Academic, San Diego p 683
25. Buckwalter JA, Glimcher MJ, Cooper RR, Recker R (1996) J Bone Joint Surg 77A: 1256
26. Athanasiou KA, Zhu C, Lanctot DR, Agrawal CM, Wang X (2000) Tissue Eng 6:361
27. Sikavitsas VI, Temenoff JS, Mikos AG (2001) Biomaterials 22:2581
28. Taton TA (2001) Nature 412:491
29. Rho JY, Kuhn-Spearing L, Zioupos P (1998) Med Eng Phys 20:92
30. Weiner S, Traub W (1992) FASEB J 6:879
31. Holmgren SK, Taylor KM, Bretscher LE, Raines RT (1998) Nature 392:666

32. Thompson JB, Kindt JH, Drake B, Hansma HG, Morse DE, Hansma PK (2001) Nature 414:773
33. Jilka RL (2003) Med Pediatr Oncol 41:182
34. Parfitt AM (1984) Calcif Tissue Int 36 [Suppl 1]:S37
35. Goldstein SA (2002) Ann N Y Acad Sci 961:183
36. Werntz JR, Lane JM, Burstein AH, Justin R, Klein R, Tomin E (1996) J Orthop Res 14:85
37. Connolly JF, Guse R, Tiedeman J, Dehne R (1991) Clin Orthop 266:259
38. Haynesworth SE, Goshima J, Goldberg VM, Caplan AI (1992) Bone 13:81
39. Bruder SP, Jaiswal N, Haynesworth SE (1997) J Cell Biochem 64:278
40. Richards M, Huibregtse BA, Caplan AI, Goulet JA, Goldstein SA (1999) J Orthop Res 17:900
41. Kadiyala S, Jaiswal N, Bruder SP (1997) Tissue Eng 3:173
42. Bruder SP, Kraus KH, Goldberg VM, Kadiyala S (1998) J Bone Joint Surg Am 80:985
43. Pittenger MF, Mackay AM, Beck SC, Jaiswal RK, Douglas R, Mosca JD, Moorman MA, Simonetti DW, Craig S, Marshak DR (1999) Science 284:143
44. Temenoff JS, Park H, Jabbari E, Conway DE, Sheffield TL, Ambrose CG, Mikos AG (2004) Biomacromolecules 5:5
45. Yoshikawa T, Ohgushi H, Tamai S (1996) J Biomed Mater Res 32:481
46. Breitbart AS, Grande DA, Kessler R, Ryaby JT, Fitzsimmons RJ, Grant RT (1998) Plast Reconstr Surg 101:567
47. Sikavitsas VI, Bancroft GN, Mikos AG (2002) J Biomed Mater Res 62:136
48. Sodian R, Lemke T, Loebe M, Hoerstrup SP, Potapov EV, Hausmann H, Meyer R, Hetzer R (2001) J Biomed Mater Res 58:401
49. Freed LE, Vunjak-Novakovic G (2000) Tissue engineering bioreactors. In: Lanza RP, Langer R, Vacanti J (eds) Principles of tissue engineering. Academic, San Diego p 143
50. Bancroft GN, Sikavitsas VI, Mikos AG (2003) Tissue Eng 9:549
51. Bancroft GN, Sikavitsas VI, van den Dolder J, Sheffield TL, Ambrose CG, Jansen JA, Mikos AG (2002) Proc Natl Acad Sci U S A 99:12600
52. Mueller SM, Mizuno S, Gerstenfeld LC, Glowacki J (1999) J Bone Miner Res 14:2118
53. Sikavitsas VI, Bancroft GN, Holtorf HL, Jansen JA, Mikos AG (2003) Proc Natl Acad Sci U S A 100:14683
54. Babensee JE, McIntire LV, Mikos AG (2000) Pharm Res 17:497
55. Bostrom M, Lane JM, Tomin E, Browne M, Berberian W, Turek T, Smith J, Wozney J, Schildhauer T (1996) Clin Orthop 327:272
56. Bowen-Pope DF, Malpass TW, Foster DM, Ross R (1984) Blood 64:458
57. Drake CJ, Hungerford JE, Little CD (1998) Ann N Y Acad Sci 857:155
58. Folkman J, Shing Y (1992) J Biol Chem 267:10931
59. Hanahan D, Folkman J (1996) Cell 86:353
60. Neufeld G, Cohen T, Gengrinovitch S, Poltorak Z (1999) FASEB J 13:9
61. Orban JM, Marra KG, Hollinger JO (2002) Tissue Eng 8:529
62. Peng H, Wright V, Usas A, Gearhart B, Shen HC, Cummins J, Huard J (2002) J Clin Invest 110:751
63. Urist MR (1965) Science 150:893
64. Wozney JM, Rosen V, Celeste AJ, Mitsock LM, Whitters MJ, Kriz RW, Hewick RM, Wang EA (1988) Science 242:1528
65. Huard J, Li Y, Peng H, Fu FH (2003) J Gene Med 5:93
66. Saito N, Takaoka K (2003) Biomaterials 24:2287
67. Vehof JW, Fisher JP, Dean D, van der Waerden JP, Spauwen PH, Mikos AG, Jansen JA (2002) J Biomed Mater Res 60:241
68. Wheeler DL, Chamberland DL, Schmitt JM, Buck DC, Brekke JH, Hollinger JO, Joh SP, Suh KW (1998) J Biomed Mater Res 43:365

69. Holland TA, Tabata Y, Mikos AG (2003) J Control Release 91:299
70. Hedberg EL, Tang A, Crowther RS, Carney DH, Mikos AG (2002) J Control Release 84:137
71. Blum JS, Barry MA, Mikos AG (2003) Clin Plast Surg 30:611
72. Rutherford RB, Nussenbaum B, Krebsbach PH (2003) Drug News Perspect 16:5
73. Blum JS, Barry MA, Mikos AG, Jansen JA (2003) Hum Gene Ther 14:1689
74. Alden TD, Pittman DD, Hankins GR, Beres EJ, Engh JA, Das S, Hudson SB, Kerns KM, Kallmes DF, Helm GA (1999) Hum Gene Ther 10:2245
75. Baltzer AW, Lattermann C, Whalen JD, Wooley P, Weiss K, Grimm M, Ghivizzani SC, Robbins PD, Evans CH (2000) Gene Ther 7:734
76. Temenoff JS, Lu L, Mikos AG (2000) Bone tissue engineering using synthetic biodegradable polymer scaffolds. In: Davies JE (ed) Bone Engineering. em squared, Toronto, p 454
77. Boyan BD, Hummert TW, Dean DD, Schwartz Z (1996) Biomaterials 17:137
78. Whang K, Thomas CH, Healy KE, Nuber G (1995) Polymer 36:837
79. Timmer MD, Ambrose CG, Mikos AG (2003) Biomaterials 24:571
80. Thomson RC, Wake MC, Yaszemski MJ, Mikos AG (1995) Biopolymers 122:245
81. Yaszemski MJ, Payne RG, Hayes WC, Langer R, Mikos AG (1996) Biomaterials 17:2127
82. Frayssinet P, Gineste L, Conte P, Fages J, Rouquet N (1998) Biomaterials 19:971
83. Ikenaga M, Hardouin P, Lemaitre J, Andrianjatovo H, Flautre B (1998) J Biomed Mater Res 40:139
84. Eggli PS, Muller W, Schenk RK (1988) Clin Orthop 232:127
85. Metsger DS, Driskell TD, Paulsrud JR (1982) J Am Dent Assoc 105:1035
86. Gauthier O, Bouler JM, Weiss P, Bosco J, Daculsi G, Aguado E (1999) J Biomed Mater Res 47:28
87. Grimandi G, Weiss P, Millot F, Daculsi G (1998) J Biomed Mater Res 39:660
88. Dupraz A, Delecrin J, Moreau A, Pilet P, Passuti N (1998) J Biomed Mater Res 42:368
89. Timmer MD, Ambrose CG, Mikos AG (2003) J Biomed Mater Res 66A:811
90. Peter SJ, Miller ST, Zhu G, Yasko AW, Mikos AG (1998) J Biomed Mater Res 41:1
91. Fisher JP, Vehof JW, Dean D, van der Waerden JP, Holland TA, Mikos AG, Jansen JA (2002) J Biomed Mater Res 59:547
92. Timmer MD, Horch RA, Ambrose CG, Mikos AG (in press) J Biomater Sci, Polym Edn
93. Vogelson CT, Koide Y, Alemany LB, Barron AR (2000) Chem Mater 12:795
94. Kikuchi M, Itoh S, Ichinose S, Shinomiya K, Tanaka J (2001) Biomaterials 22:1705
95. Horch RA, Shahid N, Mistry AS, Timmer MD, Mikos AG, Barron AR (2004) Biomacromolecules 5:1990
96. Hartgerink JD, Beniash E, Stupp SI (2001) Science 294:1684
97. Shin H, Jo S, Mikos AG (2003) Biomaterials 24:4353
98. Shin H, Jo S, Mikos AG (2002) J Biomed Mater Res 61:169
99. Massia SP, Hubbell JA (1991) J Cell Biol 114:1089
100. Rezania A, Healy KE (1999) Biotechnol Prog 15:19
101. Sofia S, McCarthy MB, Gronowicz G, Kaplan DL (2001) J Biomed Mater Res 54:139
102. Yang XB, Roach HI, Clarke NM, Howdle SM, Quirk R, Shakesheff KM, Oreffo RO (2001) Bone 29:523
103. Schaffner P, Dard MM (2003) Cell Mol Life Sci 60:119
104. Shin H, Zygourakis K, Farach-Carson MC, Yaszemski MJ, Mikos AG (2004) Biomaterials 25:895
105. McKee MD, Nanci A (1996) Microsc Res Tech 33:141

Received: February 2004

Adv Biochem Engin/Biotechnol (2005) 94: 23–41
DOI 10.1007/b99998

Simultaneous In Vivo Regeneration of Neodermis, Epidermis, and Basement Membrane

Charles E. Butler[1] (✉) · Dennis P. Orgill[2]

[1] Department of Plastic Surgery, Unit 443, The University of Texas M. D. Anderson Cancer Center, 1515 Holcombe Blvd., Houston, TX 77030, USA
cbutler@mdanderson.org

[2] Division of Plastic Surgery, Brigham and Women's Hospital, 75 Francis St., Boston, MA 02115, USA
dorgill@partners.org

Abstract Full-thickness skin loss does not undergo complete spontaneous regeneration in mammals. To restore the normal function of skin, dermal and epidermal components can be supplied by grafting various substrates, concurrently or in stages. A tissue-engineering technique that combines disaggregated autologous keratinocytes and a highly porous, acellular collagen–glycosaminoglycan matrix has been shown in a porcine model to regenerate a dermis and epidermis in vivo. During regeneration, a basement membrane of normal appearance forms at the dermoepidermal junction, and vascularization of the construct occurs. Cell-seeded grafts can be produced with either uncultured or cultured keratinocytes and can be immediately applied in a single grafting procedure. The seeding process itself and the cell culture, when used, markedly expand the donor epithelial surface area, allowing large skin defects to be repaired using grafts created from very little donor tissue. This skin-substitute technology may have useful clinical applications.

Keywords Skin · Collagen · Tissue engineering · Wound healing · Skin transplantation

1 Introduction

The dermal and epidermal components of skin can be applied to a full-thickness wound as one unit, such as a skin graft, or as separate layers. The application of separate dermal and epidermal layers is frequently performed in staged procedures to allow the grafted dermal component to become vascularized for support of the epidermis. Numerous techniques have been used for dermal and epidermal replacement. Clinically, de-epithelialized cadaveric allodermis [1], decellularized allodermis [2–4] (Alloderm, LifeCell, Branchburg, NJ), and acellular collagen-glycosaminoglycan (CG) matrix dermal analogues [5, 6] can successfully provide a dermal layer. Epidermis has been supplied with autologous skin grafts and autologous keratinocytes in the form of cultured epithelial autograft (CEA) sheet grafts [7, 8] (Epicel, Genzyme, Cambridge, MA). Numerous "composite" skin substitutes have been developed and used experimentally to provide dermis and epidermis together as one unit and in a single application step. These composites are often composed of a biodegradable construct embedded with keratinocytes and dermal fibroblasts and usually require considerable additional time in vitro to produce. Unfortunately, there are no commercially available composite grafts to permanently replace skin. Formation of an intact basement membrane (BM) at the dermoepidermal junction (DEJ) is critical for successful, permanent, functional bi-layer skin replacement. Insufficient BM development has been observed in a number of skin-substitution strategies, resulting in unstable epithelial coverage.

Immediate grafting of keratinocyte-seeded CG matrix dermal analogues allows for in vivo, simultaneous regeneration of dermis, epidermis, and BM during graft vascularization. This technique utilizes a single-stage grafting procedure and has the potential for massive epithelial surface area expansion. When uncultured keratinocytes are used, the in vitro cell-culturing is not required; thus, donor skin can be harvested and grafts prepared and grafted within hours. The simultaneous regeneration of skin components results in a functional skin replacement with dermal and epidermal layers, BM, and microvascular architecture similar to those of normal skin. This technique may offer promise for developing a clinically applicable, efficient, permanent, and functional skin substitute.

2
Skin Grafts for Full-Thickness Wounds

Skin is one of the most vital structures for mammalian survival. It functions as an environmental barrier to physical trauma, thermal injury, and infection. In humans, its approximately 1.75 m^2 surface area regulates vapor and heat exchange and protects against dehydration [9]. Massive skin loss, such as that from large surface area burns, surgical resection, or injury, is a fatal condition unless skin deficits can be replaced. Traditionally, this has been done with a surgical procedure in which areas of partial-thickness skin are transplanted from unwounded "donor site" skin to the skin defect. The use of these split-thickness skin grafts (STSGs) transfers the entire epidermis and the superficial portion of the dermis. The donor site heals spontaneously over 1–3 weeks as epithelial cells from the remnant adnexal structures (hair follicles and sweat glands) and wound edge proliferate to re-epithelialize it.

STSGs can be meshed and their surface area expanded to resurface skin defects, generally 1.5–3 times the original surface area of the STSG donor site; however, the interstices of the meshed grafts heal largely by contraction, resulting in severely compromised functional and aesthetic results with scarring, contracture, and a conspicuous meshed pattern appearance.

With skin loss involving a large surface area, the donor-site area available for graft harvest is limited. Although skin grafts can be serially reharvested several times from the same donor site, the donor site must heal prior to reharvest, thus delaying closure of the large skin defects. Furthermore, serial reharvesting progressively removes more dermis from the donor site, requiring additional time to heal and resulting in more scarring and contraction. The entire process of resurfacing large wounds from limited donor sites requires considerable amounts of time, during which the patient is at increased risk for metabolic, nutritional, and infectious complications. Furthermore, the donor sites can cause considerable patient morbidity, including pain, scarring, and hypo- or hyperpigmentation and can be aesthetically displeasing.

Skin grafts can be taken at various thicknesses ranging from those that include epidermis and only a minute portion of the dermis (STSGs) to a full-thickness skin graft (FTSG), which includes the entire epidermis and dermis. In general, the amount of scarring that occurs at the skin graft recipient site is inversely proportional to the amount of dermis included with the graft [9]. Accordingly, the thickness of dermis that is removed from the donor site is directly related to donor-site scarring. Harvesting of a FTSG leaves the donor site devoid of remnant adnexal structures to re-epithelialize the resulting wound and usually requires closure with skin advancement or a STSG from a different area.

The process of skin graft attachment and subsequent survival is referred to as graft "take", and the exact mechanism is not completely understood. It involves initial adherence to the defect surface, primarily involving a weak fibrin bond. Skin grafts are originally nourished by serum imbibition; eventually they become revascularized from the underlying tissue bed. During the 5–10 days immediately following application, the skin graft is particularly susceptible to "failure," when it fails to survive and become vascularized and integrated into the wound. Failure is frequently the result of "shear", (where the newly forming vascular channels are mechanically disrupted), graft infection, or a fluid (seroma) or blood (hematoma) collection between the graft and the underlying wound bed that prevents revascularization. If small, focal areas of the graft do not take, the underlying wound area will often contract and reepithelialize over time from the surrounding intact skin graft and wound edges. However, regrafting is usually necessary if a large surface area of the graft does not take, adding to patient morbidity with additional donor sites.

The ability to minimize or eliminate the skin graft donor-site area would be a tremendous advantage for patients with skin loss and would result in less morbidity. Unfortunately, allo- and xenotransplantated skin is not permanent and is eventually rejected, particularly the highly antigenic epidermal component. Nor is there any entirely acellular material that, when used alone for moderate or large full-thickness wounds, regenerates both epidermis and dermis. The field of tissue engineering, however, has made remarkable progress and advances in skin substitutes and artificial skin technology.

3 Roles of the BM

Arguably the most critical element for achieving permanent, functional skin replacement is the formation of an intact BM. Skin grafts contain both epidermis and dermis; intact BM, therefore, is transplanted to the wound site and maintains a physiologically normal epidermal–dermal adherence. Tissue-engineered skin substitutes depend on the reformation of a BM regardless of whether the dermal and epidermal components are supplied simultaneously or in stages.

The BM of skin is located at the DEJ, between the stratified epithelium and the dermis. The BM not only provides adhesion but also plays a critical role in interaction and signaling between these two biologically distinct tissue layers [10]. The BM also helps regulate keratinocyte proliferation, differentiation, migration polarity, and stratification.

The BM plays a pivotal role in epithelial–mesenchymal communication and prevents epidermis from delaminating from the dermis [11]. Abnormal development of BM components is associated with severe skin disease [12]. For example, anchoring fibrils are deficient in structure and/or number in patients with epidermolysis bullosa dystrophica, an inherited skin disorder in which the epidermis readily separates from the dermis and forms blisters as a result of only minimal trauma.

The BM is composed of a complex network of interconnecting structures that form a series of anchoring complexes to provide mechanical stability

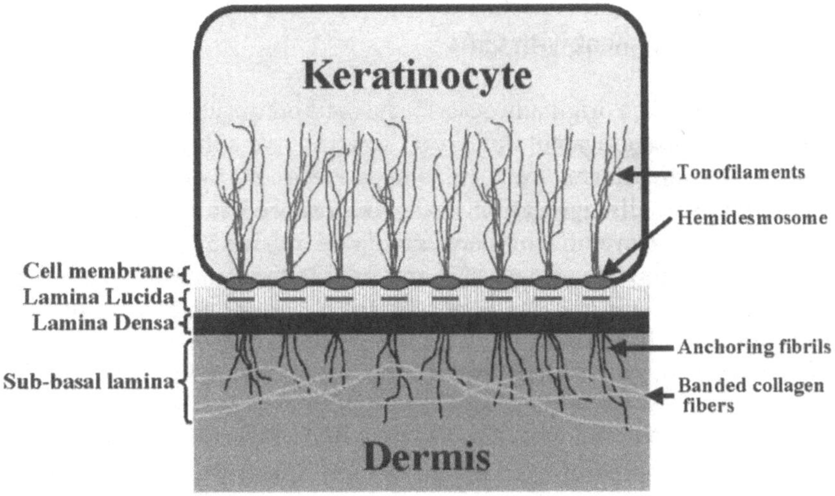

Fig. 1 Diagram of skin basement membrane with interconnecting components responsible for adherence of the epidermis to the dermis. Copyright 2002, Charles E. Butler

(Fig. 1). The BM can be divided into four distinct zones: the cell membrane, lamina lucida, lamina densa, and subbasal lamina [10]. The cell membrane contains hemidesmosomes, which consist of plectin, bullous pemphigoid antigens and $\alpha6\beta4$ integrin. Hemidesmosomes link intracellular cytoskeletal keratin fibers (tonofilaments) to laminin 5 in the underlying lamina lucida. The lamina lucida, ultrastructurally identified by an electron-lucent region on electron microscopy, contains anchoring filaments, composed of laminin 5, traversing through it. The lamina densa contains type IV collagen and appears on electron microscopy as an electron-dense area. In this zone, type VII collagen is attached to laminin 5. Finally, the subbasal lamina contains anchoring fibrils composed mainly of type VII collagen. These fibrils anchor the BM to the extracellular matrix components of the underlying dermis, which include banded collagen fibers (type I, III, and V collagen), elastic microfibrils (fibrillins), and beaded microfilaments (type VI collagen)[10]. The complex interconnections between and within these zones serve to anchor the basal keratinocytes of the epidermis to the dermis.

In both fetal development and adult wound repair, the BM elements are derived from a combination of epidermal and dermal sources, and formation of a complete, normal BM requires epidermal–mesenchymal (dermal) interaction [12, 13].

4
Epidermal Replacement with CEAs

Rheinwald and Green originally described a method of culturing keratinocytes in vitro from disaggregated single-cell keratinoctye solutions in 1975 [14]. Keratinocytes were isolated from skin biopsy specimens by mechanical and enzymatic (trypsin) disaggregation and were cultured in plastic polystyrene flasks containing a layer of confluent, irradiated murine 3T3 feeder fibroblasts to facilitate keratinocyte growth [15] and inhibit concomitant human fibroblast growth [16]. Keratinocytes were cultured in a medium containing bovine serum and growth substances including epidermal growth factor, insulin, cholera toxin, hydrocortisone, triiodothyronine, and transferrin. Subconfluent keratinocyte cultures were then serially subcultured to expand the original surface area up to 10,000-fold in 2–3 weeks [17]. At the final stage, the enzyme dispase was used to separate the postconfluent, multilayered, stratified keratinocyte sheets from the flasks without digesting the intercellular connections. These CEAs were first used by O'Connor and Mulliken [7] in 1981 for wound coverage. CEAs have been commercially available in the United States since 1988 under the trade name Epicel [18].

CEAs have the potential to provide permanent epithelial coverage from a small donor site owing to their enormous epithelial surface area expansion factor. Disadvantages of CEAs include the additional time in cell culture for preparation, susceptibility to infection, difficulties in handling the small, fragile grafts, incomplete and inconsistent take-rate of the grafts on human wounds, and high expense [19]. The most critical limitation of CEAs is the lack of a dermal layer, which can lead to long-term instability of the resulting epidermal coverage, with blistering and ulceration [20]. Insufficient epidermal adherence has been shown to be related to the absence of rete ridges and inadequate anchoring fibril formation [20].

Because the unstable attachment of CEAs to the wound bed has been attributed to the lack of a dermal substrate [21], human cadaveric allogeneic skin, with the epidermis subsequently removed, has been used as a dermal substrate for CEAs, improving the take-rate and durability to some degree [1, 22, 23]. Mesenchymal tissue has been shown to influence growth and differentiation of epithelium and improve engraftment of CEAs in animal and human studies [23–27].

5
Acellular Dermal Substrates with Epidermal Grafts

Commercially available acellular dermal analogues have also been used to provide a substrate onto which STSGs or CEAs can be grafted. For example, human decellularized allodermis (Alloderm), pre-grafted to wounds and allowed to vascularize is currently being used in conjunction with CEAs to provide dermis and epidermis to full-thickness skin defects [28]. In a porcine study, an acellular CG matrix dermal analogue was shown to markedly improve CEA take-rate when pre-grafted to full-thickness wounds and allowed to vascularize [24]. The CEA take-rate was over 98% at 7 days, demonstrating that a synthetic dermal analogue can be successfully used in combination with a CEA. The newly formed epidermis was less fragile and more resistant to separation when the CEA was grafted to vascularized CG matrix rather than control full-thickness wounds. Although not completely mature, a BM was formed by 7 days after CEA grafting. Structural analysis with electron microscopy demonstrated a lamina lucida, lamina densa and anchoring fibrils, and immunolabeling identified laminin, $\alpha6\beta4$ integrin, and type VII collagen at the DEJ [24].

6 Composite Grafts for Skin Replacement

In an attempt to provide dermal and epidermal components together in one grafting stage, numerous investigators have produced bilayer composite grafts [29–36]. These composites are usually fabricated in vitro by first forming a dermal component from the combination of a biodegradable construct with autologous cultured fibroblasts and then applying cultured autologous keratinocytes and allowing them to attach, proliferate, stratify, and differentiate. Although these grafts contain a dermal support layer, the epithelial expansion is typically far less than, and graft production time greater than, those of comparable sized CEAs [37]. Despite numerous experimental strategies evaluated, there are currently no commercially available composite grafts that can provide permanent autologous skin replacement for full-thickness wounds.

7 Dermal Regeneration with CG Matrix

Since the original use of de-epithelialized cadaveric allografts to provide a dermal substrate onto which epidermis can be grafted, a small number of commercially available acellular dermal analogues have been used clinically for dermal replacement, including Integra artificial skin.

Integra, originally developed by Yannas and coworkers, is composed of a bovine type I collagen and the glycosaminoglycan chondroitin-6-sulfate [38–43]. The co-precipitate is lyophilized and subjected to dehydrothermal treatment, forming a highly porous matrix [5]. Additional collagen cross-linking is achieved by exposure to glutaraldehyde. A silicone layer is applied to the surface and functions as a temporary epidermis to prevent trauma, dehydration, and bacterial contamination. The physicochemical properties of this matrix are strictly controlled to maintain its biologic activity in vivo [34, 38]. Some of these properties include degradation rate, mean pore diameter, glycosaminoglycan content, and molecular weight between collagen cross-links.

Integra induces cellular infiltration and neovascularization during matrix degradation, forming a well-vascularized neodermis similar to normal dermis, but without adnexal structures. The inhibition of wound contraction has been demonstrated in both rodent and porcine models [44]. The silicone layer is easily removed after 2–3 weeks and replaced with permanent autologous epidermis. This is usually a thin autologous STSG that includes the epidermis and a very small amount of dermis. Integra has been used clinically for dermal coverage since it was approved for use in burn patients in 1997 [45]. The quality of the resulting skin using this two-stage grafting technique has been shown to be equal to or better than traditional skin grafts in matched excisional burn wounds [6]. The advantages of Integra include a long shelf life with immediate availability, no risk of pathogen transmission, and the presence of a temporary silicone barrier layer. However, a second operative procedure is required to supply a permanent epidermal layer, and limited donor site surface area expansion is achieved when STSGs, rather than CEA, are used for epidermal coverage.

8 Keratinocyte-Seeded CG Matrices

8.1 In Vivo Dermis, Epidermis, and BM Regeneration

An epidermal and dermal layer can be formed in vitro with composite grafts; however, the viability of the cellular elements of such grafts must be maintained between the time the grafts are transferred from cell culture to the wound and the time at which the grafts become vascularized. Initially, nutrients are supplied and metabolic byproducts removed by diffusion through the wound fluid and the graft itself. The nutrient demand is proportional to the cellular density and the metabolic activity of the cells. Therefore, until composite grafts become vascularized, the balance between nutrient supply and demand determine cell survival, growth, and proliferation, particularly when metabolic demand is high and/or supply is limited.

Skin substitutes with nonconfluent, rather than confluent, epidermal and/or dermal cellular components have a correspondingly lower nutrient demand (Butler C, Autologous Keratinocyte Derived Skin Substitutes. Presented at the 2002 Plastic Surgery Research Council Annual Meeting, Boston, MA). In addition, the diffusion through these grafts is facilitated since their porosity is maintained by a lower cellular density and quantity of deposited extracellular matrix in the channels and pores of the scaffold material. Nonconfluent cellular composite grafts, therefore, theoretically have a reduced risk of cellular growth inhibition and necrosis.

Since the proliferation of nonconfluent cells can proceed without contact inhibition, they can proliferate rapidly in vivo, while the graft is becoming vascularized. Simultaneous in vivo (rather than in vitro) dermal and epidermal formation, particularly during revascularization, is believed to have beneficial effects on the formation of dermis, epidermis, and BM through epidermal–mesenchymal interactions (Butler C, Artificial Skin and Tissue Engineering Symposium; Strategies of Artificial Skin Development. Presented at the 2001 American Society of Plastic Surgeons Annual Meeting, Orlando, FL).

A novel skin substitute, developed in a porcine model, similar to the previously described rodent model [41], involves seeding autologous uncultured keratinocytes into a CG matrix immediately prior to grafting [46, 47]. The cells are then centrifuged through the pores and interstices of the matrix, traversing the thickness of the graft to the silicone–matrix interface (Fig. 2). The neodermis that forms is identical to that with unseeded CG matrix grafts. The seeded cells attach to the matrix fibers and individual keratinocytes and undergo clonal growth, resulting in numerous colonies that subsequently coalesce within the matrix to form a confluent epidermis that is hyperplastic and parakeratotic. The multilayered epidermis matures at between 2 and 4 weeks with continued stratification and development of a stratum corneum (Fig. 3). Keratinocyte cysts, consisting of concentric layers of stratified epidermis with a basal layer at the circumference and stratum corneum in the center, form within the neodermis (Fig. 4). Between 14 and 21 days, these structures unidirectionally migrate toward and subsequently fuse with the developing epidermis. After 21 days, the cysts have all migrated and fused with the epidermis and are no longer seen in the neodermis. When the keratinocyte cysts fuse with the developing epidermis, keratin granules contained within the cysts are expelled onto the surface of the epidermis. These keratin granules can be seen grossly as white specks on the epidermal surface that spontaneously dislodge. Keratin cysts form from the small fraction of cells unable to completely traverse the thickness of the CG matrix during the centrifugation process and are located below the silicone-matrix interface.

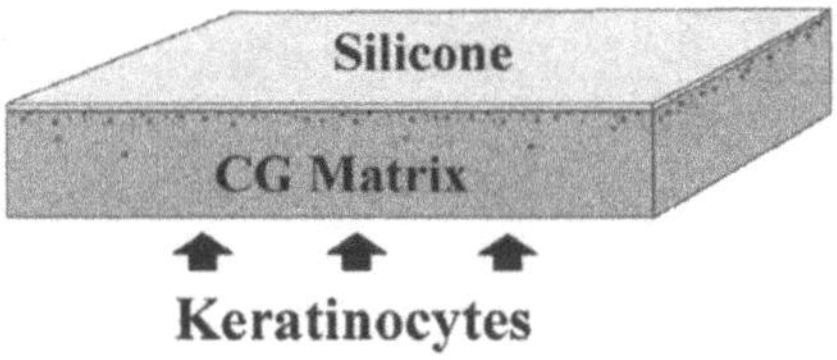

Fig. 2 Diagram of seeded collagen-glycosaminoglycan (*CG*) graft. Disaggregated keratinocytes were seeded onto the porous side of the CG matrix graft and then centrifuged through the pores and interstices of the matrix to align along the silicone-matrix. Copyright 2002, Charles E. Butler

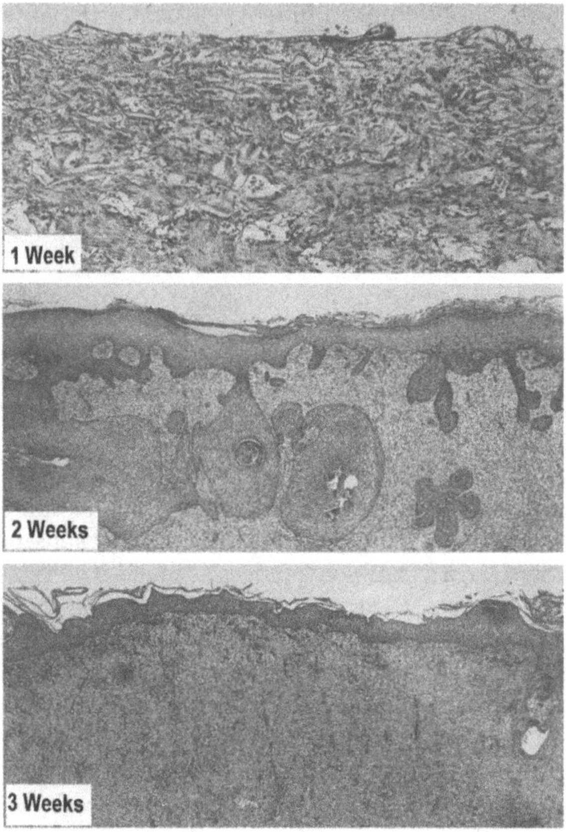

Fig. 3 Hematoxylin-eosin-stained histologic cross-sections of keratinocyte-seeded CG matrix grafts between 1 and 3 weeks. *Top* At 1 week, most of the CG matrix remains undegraded, and cellular infiltration and neovascularization of the graft have occurred from the wound base. The seeded keratinocytes have attached to the matrix and begun to proliferate, but an epidermis is not yet visible. *Middle* At 2 weeks, the CG matrix is completely infiltrated with cells and blood vessels, and a confluent, stratified epidermis is seen. Multiple keratinocyte cysts are present in the neodermis. *Bottom* At 3 weeks, most of the CG matrix has been completely replaced with vascularized neodermis. The keratinocyte cysts have unidirectionally migrated and fused with the epidermis and are no longer present in the neodermis. The epidermis is more mature and differentiated, with a distinct stratum corneum. Copyright 2003, Charles E. Butler

8.2 Effects of Seeding Density

A dose-response effect of the original keratinocyte seeding density on epidermal formation was determined with seeding densities ranging from 0 to 3 million cells cm^{-2} [46]. At 14 days after grafting, the epithelial thickness and confluence were found to have increased with increasing seeding densities (Fig. 5).

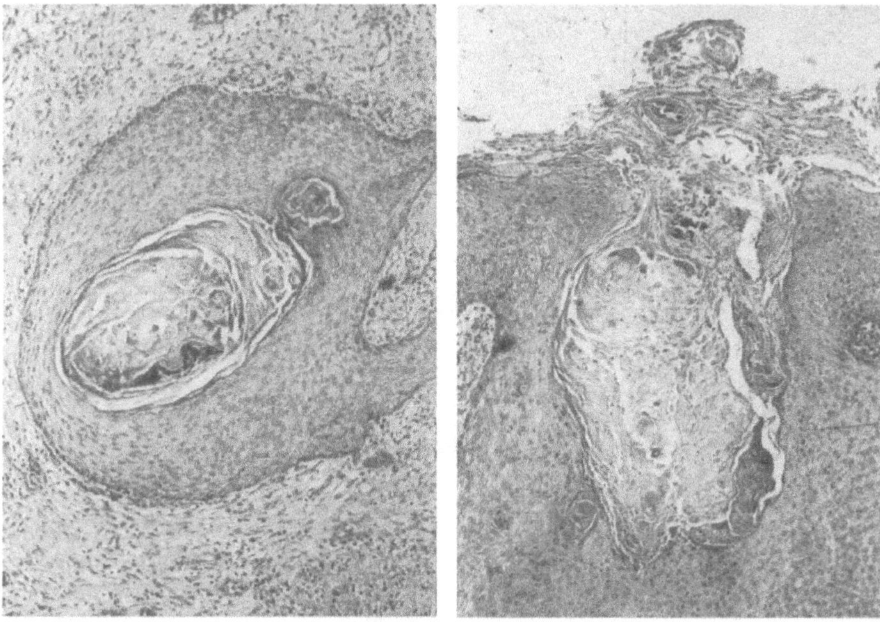

a b

Fig. 4a, b Keratinocyte cysts. **a** Keratinocyte cysts have formed in the neodermis from proliferation keratinocytes that were unable to completely traverse through the matrix to the silicone-matrix junction during centrifugation. They contain concentric layers of differentiated epidermis with a basal layer at the periphery and stratum corneum in the center. **b** Between 2 and 3 weeks, the keratinocyte cysts have unidirectionally migrated through the neodermis and fused with the developing epidermis, expelling tiny keratin granules from the epidermal surface. Copyright 2002, Charles E. Butler

There was also a direct, linear relationship between the number of keratinocyte cysts in the neodermis and the logarithm of the seeding density. The epidermis of grafts seeded with 100,000 cells cm^{-2} or more was 97% confluent, whereas the epidermis of grafts seeded with lower seeding densities was subconfluent. Seeded grafts that had subconfluent epidermis at 14 days were observed to become confluent with additional time. The silicone layer dislodged spontaneously at approximately 14 days after grafting, at which point an underlying fully confluent epidermis is desirable. A seeding density of 100,000 cells cm^{-2} was, therefore, considered to be an acceptable trade-off between graft epithelialization rate and donor epithelial surface area expansion. This seeding density resulted in an epithelial surface area expansion factor of 30-fold and consistently generated a confluent epidermis by the time the overlying silicone layer had dislodged [46]. These seeded grafts were found to have regenerated skin that was grossly and histologically similar to normal skin, but without adnexal structures, 4 weeks after grafting.

With the use of uncultured autologous keratinocytes, the cell-seeded grafts were processed and grafted within 4 h of the donor site harvest. This technique,

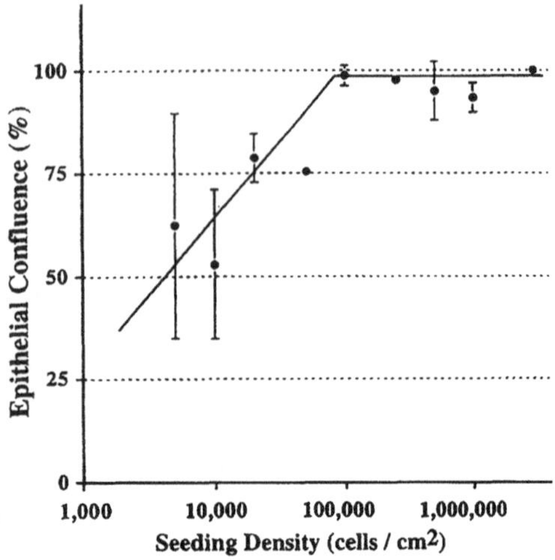
Epithelial Confluence (%)
100
75
50
25
0
1,000
10,000
100,000
1,000,000
Seeding Density (cells / cm2)

a

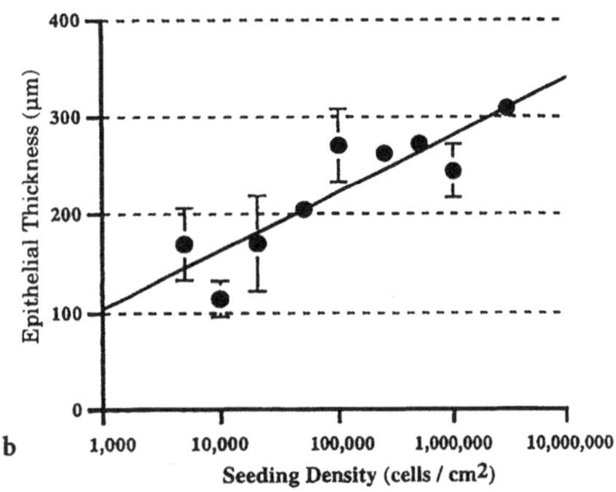
Epithelial Thickness (µm)
400
300
200
100
0
1,000
10,000
100,000
1,000,000
10,000,000
Seeding Density (cells / cm2)

b

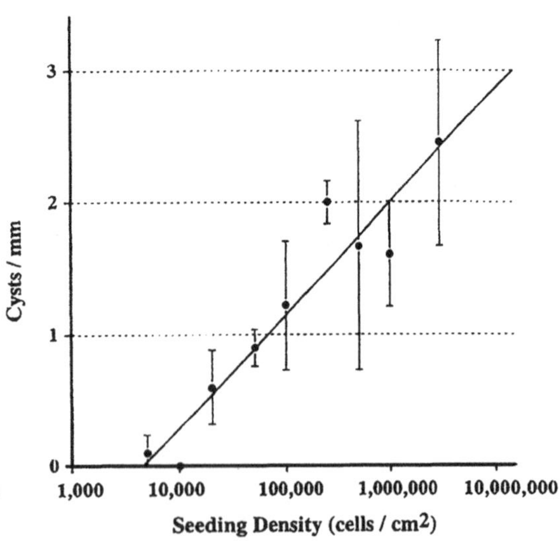
Cysts / mm
3
2
1
0
1,000
10,000
100,000
1,000,000
10,000,000
Seeding Density (cells / cm2)

c

therefore, could theoretically be used clinically with the patient undergoing grafting with the keratinocyte-seeded grafts on the same day the keratinocytes are harvested and seeded. The obvious advantage over previous skin-substitution strategies is the lack of a cell culture step and an enormous reduction in the time required for graft preparation. One disadvantage is that the epithelial surface area expansion capability is limited to that provided by the seeding process alone, in contrast to the massive expansion capabilities achieved with cell culture techniques.

8.3 Effects of Cultured Keratinocytes Used for Seeding

The effects of culturing the keratinocytes, before graft seeding on the resulting epidermis at 14 days, were studied using the same cell-seeding technique described by Butler and coworkers [46] The cultured autologous keratinocyte-seeded grafts, containing a higher proportion of basal cells, were shown to have a greater epithelial proliferative response than that of uncultured keratinocyte-seeded grafts [48]. At 14 days after grafting, the epithelial confluence and thickness, and keratinocyte cyst density were greater in the cultured than in the uncultured keratinocyte-seeded grafts. The gross and histologic appearances of the regenerated skin 6 months after grafting were indistinguishable for cultured and uncultured keratinocyte-seeded grafts(Butler C, Artificial Skin and Tissue Engineering Symposium; Strategies of Artificial Skin Development. Presented at the 2001 American Society of Plastic Surgeons Annual Meeting, Orlando, FL). Therefore, the enhanced epithelial proliferative response and greater expansion potential of cultured keratinocyte-seeded grafts are initially beneficial for coverage of large surface area defects, but the long-term results are similar to those achieved using uncultured cell-seeded grafts. The massive epithelial surface area expansion potential afforded by the use of cultured keratinocytes must be weighed against the additional time required for processing [49]. This trade-off should be considered and individualized for the particular patient when contemplating potential clinical applications of this technology.

Fig. 5a–c Epithelial confluence (**a**) and thickness (**b**) and keratinocyte cyst density (**c**) vary directly with the logarithm of the seeding density at 14 days. Epithelial confluence increased with increasing seeding density up to 100,000 cell cm^{-2} (corresponding to a 30-fold epithelial surface area expansion), after which the grafts were nearly completely confluent. Data are means±standard deviations (*bars*) for all experiments. Reprinted with permission from Butler CE et al. (1998) [46]

8.4 Time Sequence of Dermal, Epidermal, and BM Regeneration

The temporal events of simultaneous dermal, epidermal, and BM regeneration have been determined using autologous uncultured, porcine cell-seeded CG matrix grafts, seeded at a density of 50,000 cells cm^{-2} [37] The sequence of regenerative events from 0 to 35 days after grafting was characterized using histologic and immunolabeling techniques. Grossly, a translucent epidermis, visible by day 12, was opaque at day 35. Histologically, the CG matrix was nearly completely degraded by day 19, and no residual matrix fibers were identified at day 35. The seeded keratinocytes proliferated and differentiated over time to form a mature stratified epidermis. At day 4, single or small groups of keratinocytes, randomly distributed near the surface of the graft, were identified by pancytokeratin staining but were not recognizable on hematoxylin-eosin-stained sections. These cells, however, proliferated to form islands and cords, readily visible on hematoxylin–eosin sections, by day 8. A well-organized basal cell layer, evidence of keratinocyte maturation, and a confluent layer of epidermis were observed at day 12 as keratinocyte islands and cords continued to coalesce. Keratinocyte cysts were seen between days 12 and 19 and were most abundant on day 15. At day 35, the epidermis appeared histologically similar to that of normal skin. Dermal neovascularization from the wound base and edges resulted in a vascular architecture similar to native unwounded skin. Endothelial cells, without definitive vessel formation, were identified at the base of the graft using factor VIII immunolabeling on day 4. By day 8, vessels were seen penetrating the base of the matrix and extended to the surface of the graft. By day 35, the vascular architecture showed spacial remodeling with subepithelial hairpin-loop capillary arcades interdigitating with nascent rete ridges

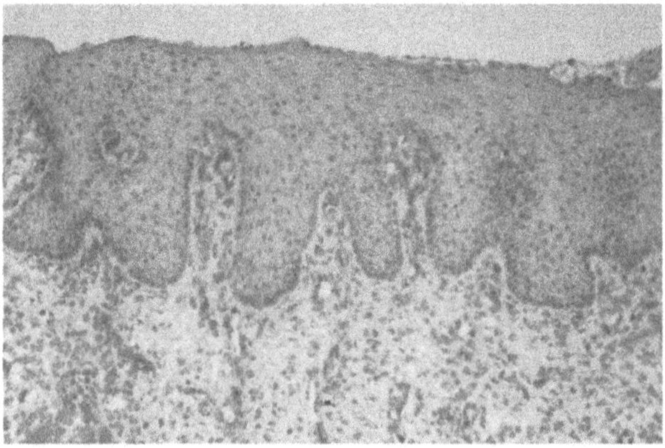

Fig. 6 Histologic cross-section at 35 days immunolabeled for Factor VIII (vascular endothelial stain) demonstrates mature vascular channels traversing through the graft with hairpin capillary loops interdigitating with the rete ridges, similar to the vascular architecture of normal skin. Reprinted with permission from Compton C et al. (1998) [37]

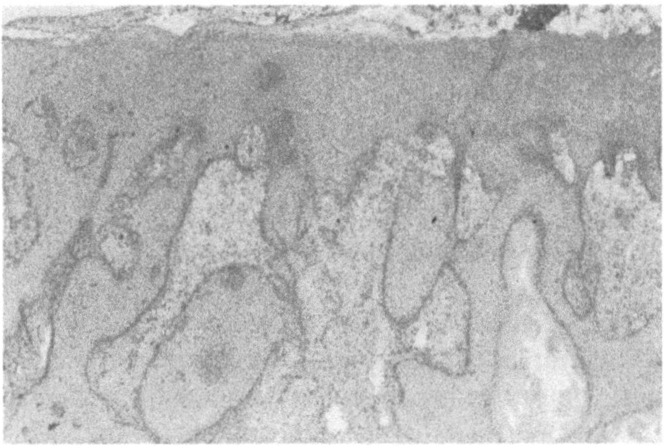

Fig. 7 Histologic cross-section at 12 days immunolabeled for type VII collagen demonstrates continuous linear staining along the dermoepidermal junction of the developing epidermis and keratinocytes cysts, demonstrating progressive formation of anchoring fibrils. Reprinted with permission from Compton C et al. (1998) [37]

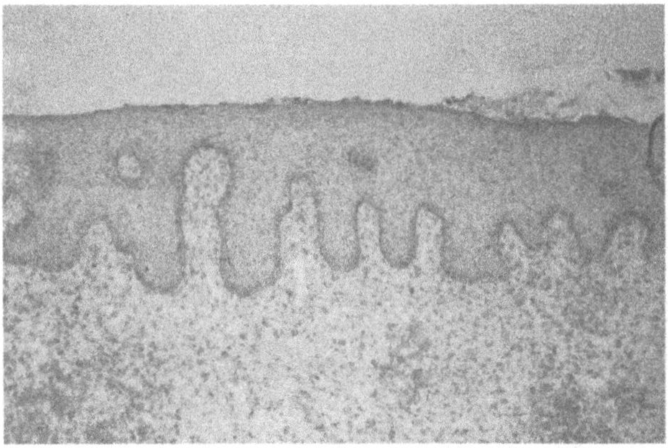

Fig. 8 Histologic cross-section at 35 days immunolabeled for α6β4 integrin demonstrates hemidesmosomes continuously aligned along the dermoepidermal junction, similar to normal skin. Reprinted with permission from Compton C et al. (1998) [37]

(Fig. 6). The BM was observed to develop and mature during the study period. BM formation occurred between days 4 and 12 with continued maturation through day 35. Immunostaining for laminin, type VII collagen, and α6β4 integrin was negative at day 4, but a discontinuous staining pattern of all three proteins was seen at the periphery of keratinocyte islands and cords at day 8. Continuous staining of these BM components was seen at day 12 along the DEJ of both the developing epidermis and the keratinocyte cysts, demonstrating maturation of the BM components including anchoring fibrils (Fig. 7). At day 35, a normal-appearing linear staining pattern of BM components was observed along the DEJ, including anchoring fibrils and hemidesmosomes (Fig. 8). The skin components reconstituted with this cell-seeded CG matrix technique were similar to those of normal skin including the dermis, epidermis, DEJ, and microvascular architecture [37].

8.5 Regeneration of Site-Specific Epidermis

This autologous cell-seeded CG matrix technology has been used to form oral mucosa by seeding human cultured oral keratinocytes into a CG matrix [50]. A 92% confluent heterotopic oral epithelium was achieved within 3 weeks of grafting of the cell-seeded grafts onto full-thickness dorsal wounds in nude mice. The regenerated epidermis appeared grossly and histologically similar to native oral mucosa (including color, thickness and stratification pattern) rather than cutaneous epidermis, demonstrating the ability to at least partially regenerate site-specific epithelium. Studies currently underway in our laboratory, at The University of Texas M. D. Anderson Cancer Center, are to characterize various site-specific heterotopically and isotopically grafted epithelia regenerated using this technique, and to compare them histologically and immunohistologically to the original donor epithelia.

8.6 Summary

The advantages of keratinocyte-seeded CG matrix technology over other skin replacement strategies include the ability to regenerate a permanent autologous epidermal layer as well as dermal layer with a single grafting procedure, massive epithelial surface area expansion capability, the potential use of uncultured and/or cultured keratinocytes for graft production, and that there is no need for in vitro culturing of cell-seeded grafts [49]. Disadvantages include the lack of adnexal structures and the need for an in vitro cell-seeding step for graft processing.

9 Discussion

Advances in cell-culturing techniques and the development of numerous biologically active scaffolds have been paramount in the development of skin substitutes. Commercially available dermal substitutes have been used clinically with excellent success, forming a permanent dermal component by remodeling, revascularization, and host cell infiltration.

Replacement of the epidermal skin component has been considerably more difficult, largely because of the antigenicity of the epidermis. Full-thickness wounds contain no adnexal structures and therefore spontaneously re-epithelialize only from the wound edge. Therefore, despite the provision of a vascularized dermal layer, only small wounds can be permanently closed, using only dermal tissue without excessive scarring and contraction, unless additional autologous epithelial elements are supplied. The most common method of providing epidermis is grafting an autologous skin graft or CEA; both techniques, however, require a second grafting procedure. Unfortunately, autologous skin grafts are limited by the size of the donor-site area, and CEAs require several weeks to produce. Furthermore, the epidermis formed from CEAs can be unstable and is prone to blistering and ulceration due to delayed, deficient, and/or incomplete BM formation.

Incorporation of nonconfluent keratinocytes into a dermal scaffold immediately before grafting is a promising strategy that may have important clinical applications. Considerable surface area expansion can be achieved by the process of dispersing (seeding) autologous keratinocytes into the dermal component. These dispersed keratinocytes attach to the matrix and proliferate, initially without contact inhibition, to form a confluent epidermis in vivo. Dispersion of individual uncultured keratinocytes from native skin, however, does not necessarily provide the optimal proliferative efficiency since only a small fraction of keratinocytes (basal cells and, to a lesser degree, transient amplifying cells) in normal skin are capable of cellular division [51]. Suprabasal cells are in various stages of terminal differentiation and are generally unable to divide [46]. Despite the low fraction of basal keratinocytes in the seeded cell population, this technique has been quite reliable.

For any technique that involves dispersion of individual, detached cells to be successful, conditions necessary for basal keratinocytes to survive and proliferate must be met; these include attachment to a substrate, delivery of adequate nutrients and oxygen, and protection from trauma, desiccation, and infection. The technique of seeding a solution of individual keratinocytes into a CG matrix dermal construct containing a pre-attached protective silicone surface layer meets these requirements [46]. The CG matrix structure allows the vast majority of keratinocytes to be centrifuged through the pores and traverse the thickness of the matrix to align at the silicone–CG matrix interface. Nutrients and metabolic byproducts diffuse through the porous matrix as neovascularization of the graft occurs.

Immunogenicity of the epidermis is a major obstacle in developing a single-stage skin substitute that can be used "off-the-shelf" without requiring harvest of any host epidermis. Systemic immunosuppression may facilitate survival of allogenic keratinocyte grafts (similar to other organ transplants); however, the side effects of immunosuppression are considerable. Strategies to reduce or eliminate donor epithelial immunogenicity without allowing tumorigenesis would be invaluable. Future developments may allow us to engineer, in vitro, a "host-specific" keratinocyte cell line using genetic engineering and/or chimeric grafts [9]. Further areas of study in skin substitute technology include recipient site-specific pigmentation, and regeneration of adnexal structures.

10 Conclusion

Keratinocyte-seeded CG matrices simultaneously regenerate permanent dermal and epidermal skin components when grafted onto full-thickness cutaneous defects in mammals. Following a single-stage grafting procedure, regeneration of skin components proceeds, in vivo, while a normal-appearing BM is formed. Both the seeding process itself and cell culture afford considerable donor epithelial surface area expansion potential. The resulting regenerated skin remains stable and is quite similar to normal skin. Further studies will be important to evaluate this technology in human clinical trials and to address the limitations of this technology, such as the lack of adnexal structures.

References

1. Cuono C, Langdon R, McGuire J (1986) Lancet 1:1123
2. Wainwright D (1995) Burns 21:243
3. Wainwright D, Madden M, Luterman A et al. (1996) J Burn Care Rehabil 17:124
4. Munster A, Smith-Meek M, Shalom A (2000) Burns 27:150
5. Burke J, Yannas I, Quinby W et al. (1981) Ann Surg 194:413
6. Heimbach D, Luterman A, BurkeJ et al. (1998) Ann Surg 208:313
7. O'Connor N, Mulliken J (1981) Lancet 1:75
8. Compton C, Gill J, Bradford et al. (1989) Lab Invest 60:600
9. Butler C, Shenaq SM, Robb GL, Miller MJ (eds) (2002) Seminars in plastic surgery, vol. 17. Thieme, New York
10. Burgson R, Christiano A (1997) Curr Opin Cell Biol 9:651
11. Fuseng N, Leigh I, Love B, Watt F (eds) (1994) The keratinocyte handbook, chap. 4. Cambridge University Press, Cambridge
12. Smith L, Sakai L, Burgeson R et al. (1998) J Invest Dermatol 90:480
13. Carver N, Novsaria H, Fryer P et al. (1993) Br J Plast Surg 46:384
14. Rheinwald J, Green H (1975) Cell 6:331
15. Rheinwald J, Green H (1975) Cell 6:317
16. Bello Y, Falabella A (2001) Am J Clin Dermatol 2:305
17. Woodley D (1989) JAMA 262:2140
18. Phillips T (1998) Arch Dermatol 134:344
19. Arons J, Wainwright D, Jordon R (1992) Surgery 111:4
20. Woodley D, Peterson H (1988) JAMA 259:2566
21. Hafemann B, Ensslen S, Erdmann C et al. (1999) Burns 25:373
22. Bergstresser H, Baxter C (1985) J Trauma 25:106
23. Compton C, Hickerson W, Nadire K et al. (1993) J Burn Care Rehabil 14:653
24. Orgill D, Butler C, Regan J et al. (1998) Plast Reconstr Surg 102:423
25. Hickerson W, Compton C, Fletchall S et al. (1994) Burns [Suppl. 1]:52
26. Kangesu T, Navsaria H, Marek S et al. (1993) Br J Plast Surg 46:401
27. Odessey R, (1992) J Burn Rehabil 13:174
28. Kraut J, Eckhardt A, Patton M et al. (1995) Wounds 7:137
29. Cuono C, Langdon R, Burchall N et al. (1987) Plast Reconstr Surg 80:626
30. Hansbrough J, Boyce S, Cooper M et al. (1989) JAMA 262:2125

31. Boyce S, Hansbrough J (1988) Surg 103:421
32. Cooper M, Hansbrough J (1991) Surg 109:198
33. Ben-Basset H, Bldad A, Chaovat M et al. (1992) Plast Reconstr Surg 91:632
34. Cooper M, Andrea C, Hansbrough J et al. (1993) J Invest Dermatol 101:811
35. Matougkova E, Vogtova D, Konigova R (1993) Burns 19:118
36. Cooper M, Hansbrough J, Spielvogel R et al. (1991) Biomaterials 12:243
37. Compton C, Butler C, Yannas I et al. (1998) J Invest Dermatol 110:908
38. Yannas I, Burke J (1980) J Biomed Mater Res 14:65
39. Yannas I, Burke J, Gordon P et al. (1980) J Biomed Mater Res 14:107
40. Yannas I, Burke J, Warpekoski M et al. (1981) Trans Am Soc Artif Organs 27:19
41. Yannas I, Burke J, Orgill D et al. (1982) Science 215:174
42. Yannas I, Lee E, Orgill D et al. (1989) Proc Natl Acad Sci USA 86:933
43. Murphy G, Orgill D, Yannas I (1990) Lab Invest 63:305
44. Orgill D, Butler C, Regan J (1996) Wounds 8:151
45. Schulz J, Tompkins R (2000) Annu Rev Med 51:231
46. Butler C, Orgill D, Compton C et al. (1998) Plast Reconstr Surg 101:1572
47. Butler C, Orgill D, Compton C et al. (1996) Surg Forum 47:752
48. Butler C, Orgill D, Correia C et al. (1999) Br J Plast Surg 52:127
49. Butler C, Navarro F, Park C et al. (2002) Ann Plast Surg 48:298
50. Butler C, Orgill D (2001) Autologous keratinocytes combined with a collagen-GAG matrix. In: Horch RE (ed) Cultured human keratinocytes and tissue engineered skin substitutes. Thieme, New York, p 243
51. Jones P, Watt F (1993) Cell 73:713

Received: February 2004

Adv Biochem Engin/Biotechnol (2005) 94: 43–66
DOI 10.1007/b99999

Spinal Cord Regeneration

Poonam Verma · James Fawcett (✉)

Cambridge University Centre for Brain Repair, Forvie Site, Robinson Way, Cambridge, CB2 2PY, UK
pv209@cam.ac.uk; jf108@cam.ac.uk

Abstract Repairing the damaged spinal cord has for a long time eluded neuroscientists. Few other achievements in neuroscience would have such a tremendous impact amongst the medical profession and the public. Despite numerous efforts, no patient has yet benefited from a regeneration therapy. Recent advances in the basic science of axon regeneration have opened new doors of hope towards prevention and cure of the devastating effects of spinal cord injury (SCI), and treatments that could encourage a partial repair of spinal injury are imminent. This article reviews the impact of SCI on patients, the current reparative strategies being explored and their potential clinical applications.

Keywords Spinal cord · Acute neuroprotection · Axon regeneration · Glial scar · Plasticity

Abbreviations

AMPA	α-Amino-3-hydroxy-5-methyl-4-isoxazole propronic acid
ASIA	American Spinal Injury Association
BDNF	Brain-derived neurotrophic factor
c-Jun	An immediate early gene
CNS	Central nervous system
CSPG	Chondroitin sulphate proteoglycans
DNA	Deoxyribose nucleic acid
ECM	Extracellular matrix
ES	Embryonic stem
EtBr	Ethidium bromide
FGF-2	Fibroblast derived growth factor
GABA	γ-Amino butyric acid
GAP-43	Growth associated protein of 43 kD molecular weight
GFAP	Glial fibrillary acidic protein
HSPGs	Heparin sulphate proteoglycans
IL-10	Interleukin-10
IN-1	Antibody against myelin inhibitory substances
MAG	Myelin-associated glycoprotein
MAP	Microtubule associated protein
MPSS	Methylprednisolone sodium succinate
MRI	Magnetic resonance imaging
NASCIS	National Acute Spinal Cord Injury Study
N-CAM	Neuronal cell adhesion molecule
NG2	Neurone-Glia antigen, a chondroitin sulphate proteoglycan
NMDA	*N*-Methyl-D-aspartic acid
NT3	Neurotrophin-3
Omgp	Oligodendrocyte myelin glycoprotein
PDGF	Platelet derived growth factor
PNS	Peripheral nervous system
RNA	Ribonucleic acid
SCI	Spinal cord injury
TGF-β	Transforming growth factor-β

1 Introduction

Spinal cord injury (SCI) is a leading cause of disability in the Western World. In the USA alone, almost 10,000 individuals, predominantly young males are rendered paralysed following spinal trauma each year [1, 2]. The commonest cause of spinal injury worldwide is automobile accidents, the next commonest being gunshot injuries in the USA and sporting injuries in Europe. In less affluent countries, accidents, gunshot and wartime injuries are the major causes. Prior to the twentieth century, victims of spinal cord trauma rarely survived their injuries. The onset of the Second World War generated numerous cases of spinal trauma and this motivated several people, most notably Sir Ludwig Guttman [3] to develop those techniques of emergency care, diagnostic tests, surgical intervention, and rehabilitation, which have become the foundations on which modern care is based. These and subsequent developments permit patients with spinal injuries to experience an almost normal lifespan, but those that do survive are often severely compromised by a devastating loss of neurological function.

The attainment of an optimal treatment strategy for SCI is hindered by the complexity of the condition and the unpredictability of the patient's prognosis. Currently the best assessment of prognosis is based on the nature of the insult; whether the injury results in a complete functional transection of the cord and on the level of the injury. One third of injuries completely transect the cord, but another third have an anatomically incomplete injury with no discernible function driven by the brain below the lesion. The final third have surviving axonal pathways passing through the injury which carry some function. A small number of axons can carry a large amount of function. There are patients who have lesions that appear to transect most of the spinal cord yet have little functional loss. Indeed a number of patients are never classified as suffering a spinal injury, because after an initial period of weakness they appear to make a full neurological recovery. An MRI scan of the spinal cord of a number of these patients may reveal a cystic injury affecting a considerable proportion of the axonal tracts passing through it.

The main determinant of the degree of disability of patients with a functionally complete injury is its level. From this point of view, spinal injury may be considered as many different conditions, because the condition of a patient with a high cervical injury is so different and the degree of disability so much greater than someone with a thoracic level injury (Fig. 1). Regardless of the level of injury however, all patients suffer loss of function of their legs and loss of control of bowel, bladder and sexual function. Of these, patients usually place a much higher priority on developing a solution to their bowel, bladder and sexual problems than on walking. Patients may spend many hours a day dealing with their bowel problems, and the unreliability of their bladders together with the infections and other issues associated with frequent bladder catheterisation are a constant source of concern. Implanted stimulation devices can help considerably in some patients, but are far from providing complete control. It is now possible to obtain viable semen from many male patients, and females with cord injuries have undergone successful pregnancies, but restoration of sexual function remains a high priority for many. As the level of injury gets higher it begins to affect upper limb function. The commonest level of injury is C6/7 (Fig. 1), and these patients have loss of hand function, although they retain control of the shoulders and elbows. Lack of hand function means that it is necessary to have at least one full-time carer to help with dressing, transfers, bladder and bowel function and other domestic tasks. For these patients, restoration of hand function is probably the highest priority. As the injury becomes yet higher, all upper limb function is lost. This removes any ability to move in and out of bed, into cars and to perform household tasks. Patients now need almost full-time assistance. Even higher injuries (Fig. 1) remove control

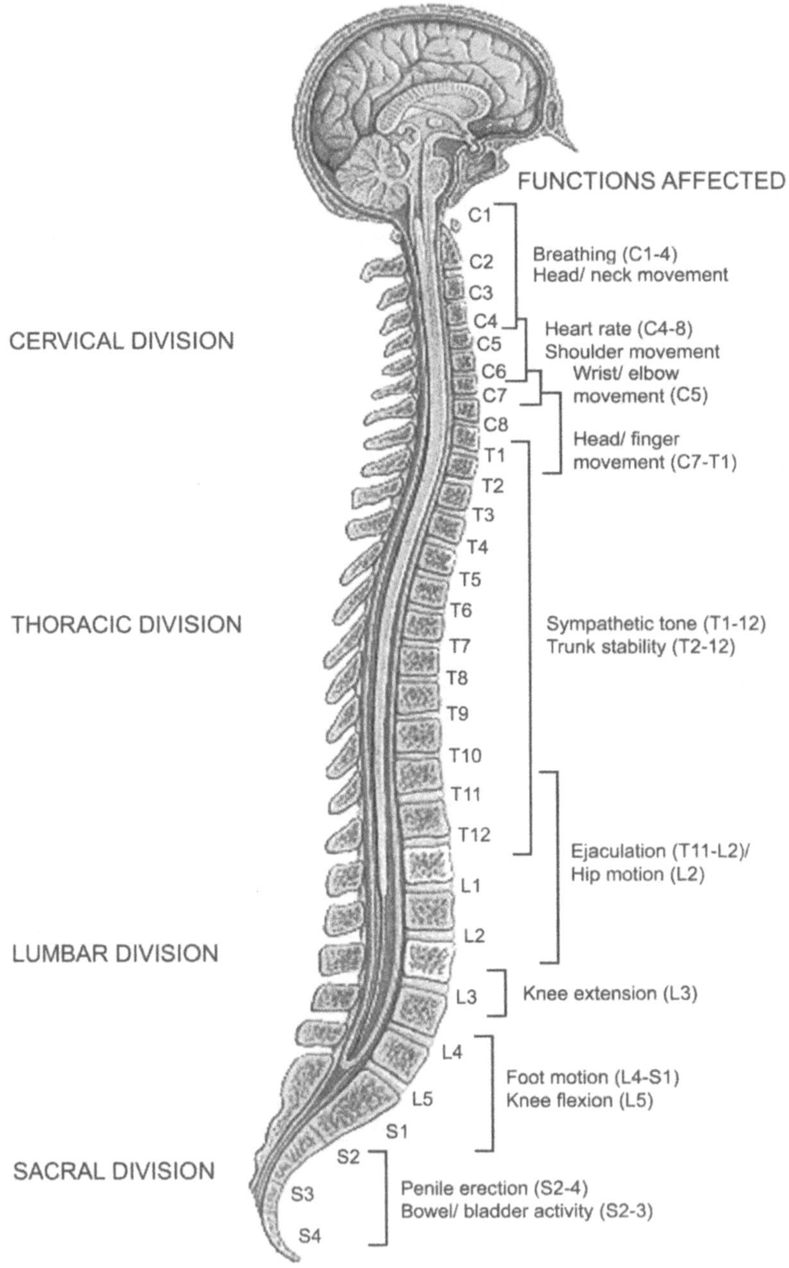

Fig. 1 The prognosis of a spinal cord injury patient is dependent upon the level of the spinal injury. Bowel, bladder and sexual function are affected by injuries of the lower segment of the spinal cord. Injuries at higher levels result in more severe presentations with the additional loss of upper limb, pulmonary and respiratory function. (Reproduced from the Spinal Research Trust, with permission)

of the diaphragm, requiring full time mechanical ventilation or a phrenic nerve stimulator, and these patients are completely dependent on a team of carers for all aspects of their life.

The discrepancy in the condition of patients with varying levels of spinal injury, an obstacle to developing a universal treatment strategy for spinal patients, provides an opportunity to those attempting to develop means to repair spinal injuries. Cajal's long-held doctrine of neurobiology suggesting the improbability of spinal cord regeneration "once development was ended, the fonts of growth and regeneration of the axons and dendrites dried up irrevocably ... Everything may die, nothing may regenerate" [4], may have been somewhat fatalistic. Although at present a complete repair of the cord remains a distant prospect, a number of treatments have been shown in experimental models to produce a partial return of function at cord levels immediately below the injury. Axon regeneration for a maximum of 4 cm has been reported following a number of treatments [5]. Each spinal level in the cervical cord is around 1 cm, so it is possible that treatments currently under development might bring about an improvement in function over two or more spinal segments. This would be invaluable in returning some diaphragm control to patients with high lesions and some degree of hand and finger control to those with C6/7 lesion. Furthermore, a more complete understanding of developmental neurobiology has provided many insights into possible ways in which neuronal regeneration in the central nervous system (CNS) may be encouraged. This research has brought us to the threshold of practical application of repairing the spinal cord following injury along three lines of approach: acute treatment, enhanced axonal regeneration or plasticity, neuronal transplantation and remyelination.

1.1 Acute Treatment

Neurotrauma is characterised by primary mechanical disruption to cord tissues from direct impact and shock waves [6–10] and a prolonged period of secondary cellular and biochemical pathophysiological alterations including haemorrhage, oedema, axonal and neuronal necrosis, metabolic perturbation, oxidative damage, release of catecholamines, excitotoxicity, inflammation and demyelination followed by cyst formation and infarction [11–17]. The developmental progression of this secondary insult is consistent with the gradual expansion of the initial lesion, both radially and longitudinally, from the epicentre of the initial force [7, 18, 19]. Thus, although the spinal cord is rarely totally transected from the initial impact, even after severe injury associated with complete paralysis [20, 21], secondary processes contribute to a complete transection-like pathology [22] and to heightened functional impairment.

The complexity and temporal nature of the secondary injury cascade affords the opportunity for pharmacological intervention on several levels [23, 24]. However, in order to carry out an optimal pharmacological intervention, the targets must be clearly identified, not only as defined static molecular and cellular factors, but as substances and receptors whose concentration and activity changes depend on the length of time from trauma and whose effects on nervous tissue probably change in a dose-dependent manner. Many pharmacological approaches aimed at attenuating the secondary injury cascade have been documented; however, in spite of optimally designed protocols affecting specific targets of the neurochemical cascade, results have been mostly disappointing. In 1990, the National Acute Spinal Cord Injury Study II (NASCIS II, USA), a multi-centre clinical trial involving 487 patients presenting with neurologically complete or incomplete lesions at intake, reported that the steroid drug, methylprednisolone sodium succinate (MPSS), when administered within 8 h of injury, improved motor and sensory recovery of patients as compared to those who received naloxone or placebo [25]. Evaluation of patients in NASCIS II at the 1-year follow-up confirmed the beneficial effects of MPSS [26]. Since the publication of the NASCIS II trial results and on the recommendations of NASCIS III, a multi-centre trial designed to determine the dose-administration protocol for MPSS, [27] the drug has been adopted as a standard of care for acute SCI in most North American centres. Although MPSS is the first accepted treatment strategy for acute SCI, criticisms regarding the drug's therapeutic physiological effects and its appreciable benefits to recipients have been raised [28, 29]. In particular, the functional neurological improvement and independence measure scores reported in NASCIS II and III respectively were only modestly enhanced in patients receiving MPSS within 8 h of SCI, suggesting that most victims are still left devastated by the physical impairments of their injury. These criticisms have led to the questioning of MPSS's therapeutic efficacy in the clinical setting and as a result, to the recently disseminated professional advice to downgrade the clinical use of this drug from that of a standard of care to an option. Nonetheless, the "therapeutic window" established by the second NASCIS trial showed that acute SCI is potentially amenable to drug therapy. Given this window of opportunity, improving drug treatment strategies aimed at attenuating early trauma-induced pathophysiology would be of great clinical value for the victims of SCI.

Since these first trials, various other potentially neuroprotective treatments have emerged from animal experiments, including control of inflammation, prevention of neurotoxicity and other interventions. Of these a GM-1 ganglioside emerged as a strong candidate for a feasible pharmacological treament of acute SCI [30]. A phase I randomised trial showed significant improvement of the American Spinal Injury Association (ASIA) motor score in patients treated within 48 h of SCI and continuing for 26 days. This preliminary report however could not be used as evidence for the routine utilisation of GM-1 in clinical practice. Although effective alone, optimal neuroprotection following SCI is likely to be achieved by a combination of pharmacological compounds directed to different pathophysiological targets; free radical scavengers [31, 32], NMDA antagonists [33], AMPA kainite antagonist and inhibitors or activators of cytokines [34, 35], taking into account the qualitative and quantitative parameters of specific targets (i.e. receptors) in the human spinal cord.

2 Enhanced Axonal Regeneration or Plasticity

2.1 Regenerating Severed Nerve Fibres

Most forms of axonal injury involve disruption to axons, which degenerate distal to the point of injury. In some injuries neurones may die but for the most part they remain preserved, although they are often atrophied. Following injury, severed axons attempt to initiate a regenerative response. In the majority of instances these attempts are aborted; regeneration is unsuccessful, and function following injury is not restored. This failure in axon regeneration is partly attributed to the extremely limited regenerative capacity of most CNS axons and partly due to the hostile glial environment of the adult CNS.

2.2 Extrinsic Mechanisms

The CNS contains many inhibitory molecules, many of which are up-regulated following injury. Several experiments investigating the regenerative capacity of sensory nerves within peripheral nervous system (PNS) and CNS environments highlight the inhibitory nature of the CNS glial environment [36]. These findings have been further supported by transplantation experiments providing various glial environments for central neurones. In a classic experiment, David and Aguayo [37] transected the spinal cords of rats and then implanted a segment of peripheral nerve across the transection, with one end inserted into the brainstem and the other into the distal stump of the spinal cord. Brainstem axons were seen to regenerate into the peripheral nerve graft and to grow the full length of the graft. However, there was little or no growth of axons out of the graft back into the host CNS, and the grafts were therefore unable to return function to the damaged cord. This study clearly demonstrated that the cells of the CNS have the capacity to regenerate over long distances when confronted by a PNS environment, but not when the same axons encounter a CNS environment.

2.2.1 The Glial Scar

Damage to the CNS rapidly leads to the formation of a dense glial scar [38], stimulated by the increased expression of glial fibrillary acidic protein (GFAP) and GFAP-mRNA, at and some distance beyond the injury site. Much more localised is the up-regulation of two other intermediate filaments, vimentin and nestin, which are found exclusively around the lesion site. The function of these later cytoskeletal changes is unknown. A functionally important astrocytic change following injury is the appearance of many junctional complexes between the cells, with the upregulation of connexion-43 expression. In the normal CNS, tight junctions are found between the astrocytic processes of the glia limitans, but after damage many of the astrocytes throughout the scar region are bound together by tight junctions. This may have the effect of making the tissue mechanically strong, but may also make it harder for axons and migrating cells to pass through it. In addition to this scar, composed primarily of hyperplastic/hypertrophic astrocytes the so-called "reactive astrocytes", there is frequently a layer of meningeal cells that migrate in from the surface and divide to form a layer around the inside of the injury site. While astrocytes are eventually the major cell type that constitutes the glial scar, early after injury there is recruitment of very large numbers of small cells with a distinctive morphology. These cells express the chondroitin sulphate proteoglycan NG2 and the PDGFα receptor on their surface. This antigenic profile and the shape of the cells, suggest that they are oligodendrocyte precursors. These cells, whose levels peak at around 10 days and then decline, form an important constituent of

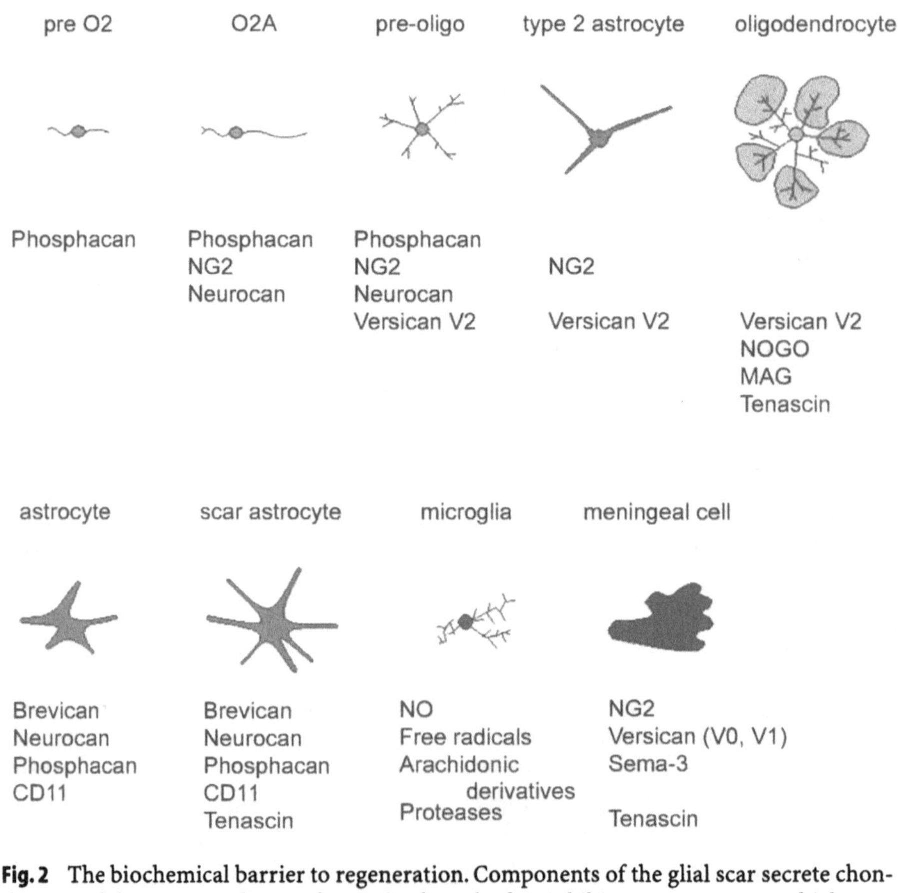

Fig. 2 The biochemical barrier to regeneration. Components of the glial scar secrete chondroitin sulphate proteoglycans, free radicals and other inhibitory components which generate a biochemical barrier to regeneration

the glial environment at the time during which axons are trying to regenerate in the CNS. The accumulation of these oligodendrocyte precursors is attributed to a combination of cell migration and cell proliferation. The migration is probably limited, since it has been shown that following demyelinating injuries, precursors are only recruited from a maximum distance of 2 mm. Much of the increase in number of this cell population must therefore be due to cell proliferation. Mitogens known to stimulate proliferation of oligodendrocytes are PDGF and FGF-2, both of which are released at CNS injuries, and so it is likely that these are the main molecules responsible for recruiting these cells.

Coupled to the physical barrier afforded by the glial scar, a biochemical barrier emerges through the cellular production of molecules such as Nogo-A, myelin-associated protein (MAG), oligodendrocyte-myelin glycoprotein (Omgp), tenascin-R, NG2, chondroitin sulphate proteoglycans (CSPGs), free radicals, nitric oxide and arachidonic acid [39, 40] (Fig. 2). Thus, the formation of a dense glial scar is one of the major impediments to neural repair [41]. Three experimental strategies have been developed to promote regenerative sprouting: blockage of glial scar formation, modification of growth-inhibitory molecules, and selective removal of reactive astrocytes.

2.2.2 Blockage of Glial Scar Formation

One of the most extreme experimental approaches to manipulating the glial environment aims to block the initial formation of the glial scar. As reactive gliosis is characterised by hyperplasia one method consists of removing all proliferating cells generated in response to CNS injury. Non-invasive low-dose X-ray irradiation has been shown to disrupt the non-permissive gliotic environment and prevent tissue degeneration around the lesion site [42] in a spinal transection model. Moreover, this reduction in scar formation is accompanied by the regeneration of numerous corticospinal axons and notable functional motor recovery [43, 44].

In addition to this non-invasive technique, attempts have been made to remove proliferating cells by injection of ethidium bromide (EtBr) into the lesion area. EtBr binds to DNA and RNA to destroy the cells through the disruption of numerous cellular functions. This treatment has been shown to be effective at clearing out CNS glia from the lesioned nigrostriatal tract following transection, producing a track through which axons can regenerate. However this treatment only opened a short window for regeneration, after which it was blocked by CNS macroglia [45]. An alternative method for the modulation of glial scar formation is to remove those cytokines that promote it, as can be achieved by injection of specific antibodies [46]. Indeed, a neutralising TGFβ-1 antibody administered into the brain of injured rats significantly attenuates the formation of the fibrous scar tissue and the limiting glial membrane [47], but this was not associated with an increase in axon regeneration [48]. In addition, the systemic administration of IL-10 (an inhibitor of microglial cytokine synthesis) reduces the inflammatory response and results in a significant attenuation of the astroglial reaction [49], which is correlated with a reduction in neuronal damage after an SCI [50].

2.2.3 Modification of Growth-Inhibitory Molecules

Since the discovery that scar formation is associated with the creation of a non-permissive biochemical environment containing specific inhibitory molecules that can block neurite outgrowth, there have been many attempts to eliminate these inhibitory factors in order to promote regenerative sprouting of the lesioned spinal cord. Numerous in vivo studies have demonstrated the inhibitory impact of myelin on axon regeneration. Early reports emerged suggesting that preventing myelin formation or using a monoclonal antibody (mAb IN-1) that neutralised the activity of the inhibitory myelin protein Nogo-A promoted regeneration of corticospinal fibres in the lesioned spinal cords of adult rats [51–53]. These observations were followed by immunisation of lesioned animals with CNS myelin, lysation of myelin with antibody and complement in vivo, or enhancement of myelin clearance by injecting activated macrophages [54–56]. Although inhibition of Nogo-A after SCI leads to some regeneration of the damaged axons and also to sprouting of unlesioned axons, it is not clear what proportion of the return of function is due to the regenerated axons and what proportion to plasticity.

Current efforts continue to focus on blocking Nogo-A, as well as other growth-inhibitory molecules by the in vivo application of antibodies against these molecules, by blocking the receptors to which they bind, or by pharmacologically manipulating the downstream signalling pathways induced by these inhibitory signals in growing neurites [57, 58]. The discovery that the Nogo-A receptor exhibits neural specificity and is also a high-affinity receptor for other known myelin inhibitory proteins (MAG and OMgp), suggests that there will be increased activity around this therapeutic target [59–61]. The convergence of three inhibitory influences (Nogo-A, MAG and OMgp) at the level of both cell surface receptors and intracellular signalling could be a reason for the devastating suppression of neuronal growth by CNS myelin after SCI. Whether blocking of Nogo-A, MAG, and OMgp will have additive effects, and produce a better effect than the increase in regeneration produced by blocking the activity of each individual growth-inhibitory protein, remains to be ascertained [52]. Despite initial findings, the exact biology of Nogo-A has been made somewhat uncertain by contradictory findings in parallel studies of knockout mice lacking Nogo-A [57, 62, 63]. The reason for the varied results is unclear, but one possibility is that minor differences in the way that the genes were inactivated or the strain background of the different animals is responsible.

Recent research has identified other attractive targets for potential therapies, namely the Nogo-A receptor itself or the intracellular signalling cascades downstream of the receptor. Grandpre et al. recently identified a competitive antagonist of the Nogo receptor, derived from amino-terminal peptide fragments of Nogo-66 [59]. They found that the Nogo-66 (1–40) antagonist peptide (NEP1–40) blocked the myelin inhibition of axonal growth in vitro. Intrathecal administration of NEP1–40 into rats with a mid-thoracic (T6–7 level) hemi-section induced significant growth of the corticospinal tract below the lesion. In addition, they observed a significant improvement in motor functioning. However, myelin inhibition is not the only factor responsible for the lack of regeneration; astrocytes and other glial cell types also produce inhibitory molecules including CSPGs and tenascin [64].

CSPGs are expressed in the CNS. Most are inhibitory to axon growth and several are up-regulated in glial scar tissue. They therefore represent a barrier to the re-growth of axons. Application of bacterial chondroitinase ABC at the site of injury digests the glycosaminoglycan side chains of these CSPGs, which are an important component of their inhibitory activity, disrupting the growth-promoting properties of the heparin-sulphate proteoglycans (HSPGs). Studies utilising the contusion model of SCI have demonstrated a considerable reduction in CSPG immunoreactivity in response to chondroitinase ABC application [65]. Furthermore, when chondroitinase ABC was placed on top of the interface between the hemisectioned spinal cord and a nerve graft, it promoted substantial axonal growth into the grafts [66]. When partial spinal transections were treated with chondroitinase it promoted both axon regeneration and functional recovery [67, 68]. Subsequent experiments in crush injury models and other species have shown the efficacy of chondroitinase in promoting both regeneration and return of function. Removal of CSPGs with chondroitinase promotes both axon regeneration and plasticity [69], and it is not yet clear which is the more important for promoting return of function.

Yet another approach to counter the inhibitory effects of the fibroblastic "lesion core" composed of meningeal and other cells, has been to interrupt the synthesis of the extracellular matrix (ECM) itself. Studies carried out by Stichel and Muller [70–72] suggests that the impermeable nature of the scar is principally due to its basal membrane formed of collagen IV and laminin. Reducing collagen IV deposition by the iron-chelator 2–2′-dipyramidine after fimbria fornix lesion prevents the formation of a dense collagen network and allows extensive regeneration of the lesioned axons in the commissural track. However, this treatment is not sufficient to enhance corticospinal regeneration after a mid-thoracic SCI [73], although a more powerful chelator may be effective [74]. Other components of the ECM, in particular the proteoglycans, can inhibit axon regeneration. One of the major astrocyte proteoglycans with inhibitory properties is neurocan. This molecule is strongly up-regulated in response to CNS injury. The properties of this molecule probably explain why three-dimensional astrocyte cultures are inhibitory while monolayers are permissive. Neurocan is secreted in large amounts by cultured astrocytes, but it does not bind to the astrocyte surface and therefore floats away from the upper surface of the cells to be diluted in the medium: it therefore has little inhibitory effect on axon growth. However, in three-dimensional cultures, neurocan is trapped between the cells and is therefore present in the environment surrounding regenerating axons [38]. In addition, meningeal cells express the axon growth inhibitory molecules NG2, versican, and semaphorins 3A and 3C [75, 76].

Modulation of the intracellular signalling molecules that regulate the inhibition of neurite outgrowth, principally cAMP levels, may be another means to achieving axonal regeneration. Previously, it was demonstrated that the central projections of dorsal root ganglia would regenerate if their peripheral projections were lesioned 1 week earlier [77], also known as the "conditioning lesion". A study by Qiu et al. reported that this conditioning lesion effect manifests itself via a cAMP mechanism and that cAMP plays an important role in promoting extensive regeneration of cut dorsal root neuronal central projections into the dorsal columns of the spinal cord [78]. An earlier study showed that elevated cAMP levels could block myelin-specific inhibitors, myelin-associated glycoproteins (MAGs) and myelin in general [79]. The concept of targeting intracellular signalling molecules extends beyond raising cyclic nucleotide levels in neurones in vivo. Many of the inhibitory effects converge on the Rho pathway, especially relevant in relation to recent studies that have identified a potential co-receptor for the Nogo-A receptor as the neurotrophin receptor, p75NTR, and a possible second-messenger pathway involving RhoA [80–84]. Blockage of RhoA with a bacterial enzyme that ribosylates has been shown to enable axons to overcome inhibitory environments in vitro and to regenerate after lesions to the optic nerve and spinal cord [85].

One major concern with this and other treatments that promote sprouting and plasticity such as chondroitinase, is that it may enable maladaptive sprouting and growth of non-injured neurones, which could disrupt the highly ordered connectivity of the CNS established during development. Indeed, the functions of Nogo-A, MAG or OMgp may be related more to their possible physiological role in the preservation of the wiring of the CNS, by locking connections into place, than to the suppression of neuronal regeneration.

2.2.4 Removal of Astrocytes

Reactive astrocytes are the major cellular constituents of the glial scar. The tightly bound processes of these reactive astrocytes may constitute a mechanical barrier, whilst their production of growth-inhibiting molecules, up-regulated following injury, induces a biochemical barrier to any neuronal repair. Attempts to limit the formation of the astrocytic scar, with 7β-hydroxycholesterol-oleate (a cholesterol derivative) [86], have proved successful not only in reducing astrocytic levels, but also in enhancing local sprouting in serotonergic axons below the lesion. These serotonergic axons establish normal synaptic contacts. Also, transgenic animals lacking the two main astrocytic intermediate filaments, GFAP and vimentin show enhanced axon regeneration after injury. Experiments with single knock-out mice deficient in GFAP or Vim are currently under way to eventually elucidate the possible contribution of each protein to the astroglial reaction and axonal sprouting of supraspinal neurones. This and future work in this sparsely investigated field has the potential to produce fruitful dividends.

2.3 Intrinsic Neuronal Control of Regeneration

Much axonal regeneration research focuses on the inhibitory effects of the CNS environment. However, the axons are not merely passive bystanders whose regenerative ability is entirely dictated by their environment. Part of the reason for the failure of regeneration can be attributed to the nature of the axon itself. Three main intrinsic properties affect axonal regeneration; neuronal age, the site of transection and axonal variability.

2.3.1 Neuronal Age

Embryonic neurones grafted into the adult CNS will often extend axons considerable distances through the inhibitory glial environment [87, 88]. The observed differences in the capacities of embryonic and adult neurones for growth could be attributed to changes in the internal machinery for axon growth, or owing to changes in the molecules by which axons respond to their environments, particularly adhesion molecules and receptors for inhibitory molecules. Modifications in molecules mediating interactions with the ECM, integrins, and those mediating cell–cell interactions, have been documented. Integrins mediate growth of axons on ECM molecules, and embryonic axons are able to adjust their cell surface integrins to suit their environment; adult axons appear to have lost this ability [89]. The adhesion molecule N-CAM has several splice variants and, after development, an axon VASE is inserted. This peptide is inhibitory to axon growth, which may partially explain the poor regeneration of mature axons that express it [90]. In addition there are differences in the molecules expressed within various axons. Growth-associated protein-43 (GAP-43) is present in all growth cones during development, and only expressed in regenerating axons [91, 92]. Microtubule-associated proteins (MAPs) stabilise the axonal cytoskeleton and are necessary for axonal growth [93], yet the type of MAPs present in axons changes late in development, and the embyronic forms are not re-expressed during regeneration [94]. Growth factors can also greatly increase the vigour of regeneration, yet their receptors are down-regulated in many axons after growth and may not return during regeneration. Finally, whilst adult axons are critically dependent on the presence of autocrine survival factors it appears that embryonic axons do not require such support, which may explain their more robust regeneration in hostile surroundings.

2.3.2
Site of Transection

In contrast to axons of the PNS, the regenerative potential of CNS axons is related to the distance of transection from the cell body. The pioneering experiment of David and Aguayo [37], placing peripheral nerve grafts into the spinal cord, demonstrated that only axons close to the graft regenerated. This poses a major problem for spinal cord repair, since many of the vital axonal pathways that need to be repaired are long: for example, the pyramidal neurones of the motor cortex. Thus, there has been considerable interest in seeking to identify the mechanisms behind this phenomenon, and in finding methods to induce regeneration in axons severed far from the cell body. GAP-43 expression correlates well with regenerative potential. The application of neurotrophic factors to severed axons results in an increased expression of GAP-43 in the cell body [95]. This treatment has been the most effective treatment to date and there are now several examples where it has been used with success in vivo [96, 97]. Brain-derived neurotrophic factor (BDNF) in particular, delivered to the cell bodies, has been found to increase the ability of axons transected far from the cell body in the rubrospinal tract [98]. Neurotrophins have also been successfully used as an adjunct to treatments to block oligodendrocyte-mediated inhibition. Schnell and Schwab investigated the effects of BDNF and neurotrophic factor-3 (NT-3) on regeneration of corticospinal axons. They found that NT-3, and to a lesser extent BDNF, injections into the cord caused a considerable increase in the sprouting of the cut axons proximal to the lesion. This treatment was then coupled to the antibody IN-1, which by itself induces some axons to regenerate after they are cut. The combination of IN-1 and NT-3 increases the density of regenerating fibres, and the maximum distance of regeneration.

2.3.3
Neuronal Type

Different types of axons vary greatly in their ability to regenerate. Perhaps the most extreme example of this is the Purkinje cells of the cerebellum whose axons appear to have no regenerative response to axotomy. At the other extreme are spinal motor neurones. These axons are probably the best axons in the CNS at regenerating, especially when cut close to the cell body. These axons have been used as the basis of a technique for repairing injuries where the central roots are avulsed from the cord: peripheral nerve tissue is re-implanted into the cord, and the motor axons regenerate into it. The reasons for these observed differences are not established. Purkinje cells do not up-regulate growth-associated proteins such as GAP-43 or c-jun after axotomy, unlike motor neurones, but transgenic animals in which these molecules are expressed in Purkinje cells do not regenerate their axons. Some up-regulation of these molecules in Purkinje cells is seen when animals are treated with Nogo-A blocking antibody, suggesting that myelination may have some part to play in suppressing regeneration [99].

2.4 Enhancing Plasticity

Although spontaneous regeneration of lesioned fibres is limited in the adult CNS, many people that suffer incomplete SCIs show significant functional recovery. This recovery process can continue many years after the injury and probably depends on the reorganisation of circuits that have been spared by the lesion. Synaptic plasticity in pre-existing "silent" pathways and the formation of new circuits through collateral sprouting of both lesioned and unlesioned fibres are important components of this recovery process. Thus, extensive functional and anatomical reorganisation has been demonstrated to occur in the spinal cord, above and below the SCI, and in centres of the brain that have some input into spinal motor pools [100].

Although pronounced before and just after birth, this plasticity (that is, the ability of one group of nerve fibres to take over the role of another injured group) is much less prominent in the adult CNS (the so-called "early lesion effect" or "Kennard principle") [101, 102]. Nevertheless, the fact that it occurs in adults at all provides the mechanism for the partial recovery of function that is seen after partial SCI, stroke, head injury and other pathologies. The processes underlying this redistribution of function are the activation and strengthening of silent synapses by feedforward and feedback inhibition and sprouting. Many neuronal connections in the CNS are kept silent and ineffective by inhibitory mechanisms, probably mediated by the neurotransmitter GABA. Following damage, many of these synapses can become active, often within minutes, allowing some redistribution of function. In addition, when axons are severed, and the synaptic connections destroyed, vacated synaptic sites become reoccupied by compensatory growth of spared, intact nerve fibres above and below the lesion. Enhancement of either or both of these processes may prove critical to the field of spinal cord repair.

Antibodies against neurite growth-inhibitory factors have been shown to enhance compensatory sprouting of nerve fibres in the partially lesioned brain stem and spinal cord of the adult rat. This resulted in an almost complete functional restoration of fine paw movements in these animals, as demonstrated by the food-pellet-reaching test [103, 104]. Several of the interventions that promote regeneration of damaged nerve fibres (neurotrophic factors, antibodies) will almost certainly also enhance compensatory growth of non-lesioned nerve fibres. In addition, functional and anatomical evidence exists that spontaneous plasticity can be potentiated by activity, as well as by specific experimental manipulations. In particular, the flow of electrical impulses in a particular neural circuit is known not only to strengthen connections, but may even induce the sprouting and formation of new connections. It is in this context that the application of electric fields may prove to be beneficial in the setting of SCI and specific rehabilitation strategies may even come into play by this means [105]. The effects of activity on the molecular machinery of nerve fibre growth are currently under investigation, as are the effects of combining growth factors with specific training programs for treating SCIs. Studies in animal models should pave the way to a better understanding of rehabilitation treatments and to the development of new therapeutic approaches for people with injured spinal cords.

3 Neuronal Transplantation and Remyelination

3.1 Bridging Cysts and Scars and Cellular Replacement

Within several weeks of injury to the spinal cord, macrophages migrating from the bloodstream have cleared the tissue debris at the lesion site, resulting in fluid-filled cysts surrounded by scar tissue. Sprouting nerve fibres do not often cross this stretch of foreign territory, in particular because scar tissue at the injury site contains scar-associated neurite growth-inhibitory molecules. Scar formation is a natural reaction of the lesioned CNS tissue, and attempts to prevent it have only partially been successful. The complex interplay between inflammatory cells and astrocytes, and the glial cells of the nervous system that are the principal contributors to scar formation, is still poorly understood. Bridges that form a growth-permissive scaffold within the lesion site should greatly facilitate regenerative axon growth across the injured area, leading to the re-introduction of axons into the remaining intact parenchyma and the return of substantial neurological function. Many types of cells, tissues, or artificial materials have been implanted as bridges into the injured spinal cord [106]. However, the success of these experiments is limited, often because astrocytes "wall off" the foreign object, thus greatly restricting access of regenerating fibres to the implants. Important exceptions to this are implants made of olfactory nerve glial cells (the only neuronal cells within the CNS in which spontaneous axonal regeneration occurs after sectioning) or precisely placed peripheral nerve grafts, which are invaded by regenerating axons and can serve as bridges across even anatomically complete lesions. Unlike Schwann cell grafts, which are walled off by astrocytes, olfactory glia align themselves with the lesion site, migrate long distances into the anterior and posterior spinal cord stumps, and guide regenerating axons through the lesion [107, 108]. Transplantation of these cells to the injured spinal cord have produced extensive axon regeneration and dramatic functional recovery. The migratory behaviour of these specialised glial cells, which are also known to produce a variety of growth factors and growth permissive ECM molecules, may be part of their success.

Several groups are currently exploring the potential use of neural stem cells [109]. Neural stem cells can be obtained by expanding a small biopsy of brain or bone marrow tissue in culture. The lengthy process of cell purification and expansion of autologous cells means that transplant techniques cannot be used in acute injuries unless cell lines or cells from donors are used. In principle, therefore, there are attractions in designing cell-free grafts of biocompatible bridging materials that use and control endogenous cell populations within the injury site in order to promote axonal regeneration and control inflammatory and glial reactions [110]. The substrate required will ideally be immuno-tolerant, will possess a porous scaffold to enable nerve regeneration and cell repopulation, will have a molecular make-up that can be easily manipulated and will be easily integrated into the CNS. Fibrin, or freeze-dried alginate, gels loaded with growth factors that attract regenerating neurites and keep scar formation under control, show much promise in this respect and are attractive options because of their reabsorbable properties [111]. In addition, rapidly absorbable substrates such as poly-α-hydroxyacids and vinyl chloride and acrylonitrile porous tubes are currently being tested in the transected spinal cord [112] with some success, although there are some concerns over mechanical hindrance and their lack of absorptive properties.

The principal obstacle to be overcome with any bridging material is how to integrate them into the spinal cord tissue, preferably non-invasively, without inducing scar formation. In this case, the replacement of scar tissue with a more attractive cellular substrate might lead to a more functional repair. Either way, the use of any bridge will have to be part of an extremely complex series of treatments involving trophic factors and factors which neutralise inhibitory molecules, all of which together will allow significant axon regeneration in patients with SCI. Although there are many hurdles to overcome, reabsorbable substrates that are integrated with cells, guidance molecules or electric fields represent an exciting new approach for spinal cord repair.

3.2 Replacing Lost Neurones

Recovery in the damaged spinal cord is hindered by the limited ability of the vertebrate CNS to replace lost cells, repair damaged myelin, regenerate long tract axons and re-establish functional neural connections. Endogenous neural progenitor cells are susceptible to many of the same injury processes as mature cells and the response of endogenous progenitor cells to nervous system injury is inadequate to replace lost tissue. Therefore, an obvious approach for cellular replacement has been transplantation.

3.2.1 Neural Transplantation

Many sources of transplantable cells have been used for the purpose of SCI repair. Mature CNS cells are post-mitotic and do not divide in culture, making it difficult to amass large numbers of cells for transplantation. Primary CNS cells can be induced to re-enter the mitotic cycle by transformation with genes from tumour cells, but transplanting cells containing cancer genes is less than ideal. Fetal CNS cells can, on the other hand be expanded in culture, will survive transplantation and will generate nerve fibres making functional synaptic connections. However, ethical considerations of using fetal tissue in research or clinical trials make the widespread use of fetal cells impracticable. Recently, a novel source of replenishable pluripotent cells for transplantation has become available through the use of embryonic stem (ES) cells.

3.2.2 Replacing Lost Neurones from Stem Cells

Stem cells are undifferentiated cells that have the capability to divide and produce many other types of cell. Several unique features of ES cells make them excellent candidates for transplantation. These features include their plasticity (pluripotentiality) and relative lack of immune excitation. Furthermore, ES cells are genetically manipulable and double allele knock-ins and knock-outs are possible in single cells [113]. The scientific power of ES cells is just beginning to be harnessed at the individual cell level. Recent demonstrations suggest that the embryo is not the only source of stem cells; evidence for stem cells in the adult brain raises the attractive possibility that endogenous neurogenesis may be manipulated to therapeutic advantage after CNS injury [114]. Not only are the cells remarkably plastic: adult hippocampal stem cells can give rise to regionally specific cell types, not only in the hippocampus, but also in the olfactory bulb, cerebellum, and retina, but they can even migrate considerable distances. Thus, neural stem cells from both embryo and adult seem to be "reprogrammable", an idea, drawn from haematopoiesis, that there is a genetically "naive" brain stem cell. What is proving truly remarkable is the range of cell types that can be produced in a culture dish when the environment, growth or trophic factors, in combination with particular substrates, is configured appropriately.

The contribution of developmental neurobiology to our understanding of neuronal regeneration and the implications for CNS cellular transplantation is clear. There may exist neuronal "stem" cells in the adult CNS that may be cultured and used for CNS repair. If transplantation of cells into the CNS is to continue to progress, lessons from the developing CNS, such as the way in which a developing cell "integrates" into its surrounding environment in a three-dimensional context and the factors which support this occurrence, must be applied.

3.3 Treating Demyelination

Many nerve fibres in the CNS are insulated by myelin, a sheath formed by glial cells called oligodendrocytes. Demyelination is a critical contributor to the neurological deficit following SCI, because it renders axons that remain anatomically intact physiologically non-functional due to a disruption of electrical impulse conductivity. Although lengths of demyelinated axon may acquire the ability to conduct by continuous or non-saltatory conduction, this form of conduction is less rapid and more vulnerable to changes in the environment, such as temperature. Moreover, some demyelinated axons do not conduct action potentials due to an imbalance between densities of sodium and potassium channels. Remyelination can occur spontaneously in the CNS and its role in recovery of function make it a particularly fascinating area. However, this remyelination is inconsistent, and in many cases fails to occur, therefore optimising remyelination through transplantation of myelin-producing cells offers a pragmatic approach to restoring meaningful neurological function. Since each oligodendrocyte myelinates 10–20 different axons, loss of a single oligodendrocyte can cause dysfunction in many different axons. Myelin damage appears to result from inflammatory processes in the injured tissue and may involve delayed apoptotic death of oligodendrocytes, the glial cells that produce myelin in the CNS [115]. The ability of the spinal cord to replace lost myelin over time is uncertain. While studies of recently injured cords show extensive demyelination, cords injured 2 years or more previously have none, and also contain areas of Schwann cell myelination. Either the cord can remyelinate damaged axons over a period of time, or demyelinated axons gradually die [116].

Two approaches are in clinical development. The first is pharmacological treatment with 4-aminopyridine, or fampridine, a potassium channel blocker that can restore action potential conduction in demyelinated, or partially myelinated, axons in the injured spinal cord [117]. Clinical studies in more than 200 patients over 5 years have reported modest beneficial effects on motor and sensory functions, significant reductions in spasticity and hyperreflexia and evoked potential evidence of enhanced conduction within spinal cord pathways [118, 119]. So far, the only significant side effect is a minimal risk of seizure, presumably because the compound also increases the excitability of healthy neurones. Large-scale human trials of the drug have already begun. The extent of benefit to an individual will be expected to depend on the particular surviving pathways within the spinal cord, as well as their state of myelination.

A more challenging, though possibly permanent method of treating demyelination may be offered by cellular transplantation of either oligodendrocytes or Schwann cells. The natural reparative response to CNS injury is inadequate for a number of reasons, and therefore transplantation of replacement oligodendrocytes is a pragmatic approach to the problem. Neural stem cells, obtained from adult animals or human brain biopsies (and in the future, possibly from adult bone marrow) and expanded in culture, could also be primed with the correct growth factor cocktail to become oligodendrocytes or Schwann cells, which could then be implanted into lesion sites to promote myelin production. Successful remyelination by such cells has been shown to occur in the spinal cord of adult rats, with restoration or enhancement in axonal conductivity [120–122]. There is also data to indicate that olfactory ensheathing glial cells enhance limited functional axonal regeneration [123]. Repair of the myelin sheath and subsequent restoration of impulse conductivity in nerve fibres that have survived the lesion would enable patients to make much better use of their surviving nerve fibres' tracts. The major issues, however, as with most other cellular therapeutic approaches revolve around source, delivery, control and immunogenicity.

4 Future Expectations

Long regarded as impossible, partial spinal-cord repair may soon become an attainable goal of clinical therapeutics. A total cure is seldom required; even a limited repair of the damaged spinal cord can produce a disproportionate return of function. In animal models approximately 10% of the functional connections across the lesion are required to support ambulation [124]. Therapeutic strategies should focus on a step-wise restoration of function. The aims should not simply be to restore motion, but should include return of crucial limited functions such as the recovery of bowel and bladder control, improved pulmonary function, and enhancement of limb movements. Based on evidence available already, some reparative therapies may prove to be more effective than others. Exciting advances are occurring in the field of neural regeneration. The current climate of spinal cord repair offers neurobiologists a rich and evolving plethora of tools. An increased understanding of the CNS development, the nature of neuronal injury, the factors which cause neuronal death and inhibit neuronal re-growth, the way in which axons direct (and redirect) themselves to a target and the role and actions of neurotrophic factors have begun to hint at ways neuronal injury can be treated more effectively clinically, and animal models are valuable assets on which to trial these strategies. Clearly, there is no point demonstrating anatomical re-connectivity if this does not produce an improvement, or even a restoration, of animal function.

In general, studies continue to show promising results that suggest it might be feasible to enhance neuronal growth by blocking glial inhibition. But even if all the inhibition could be overcome, there is still one problem: neurones must be stimulated throughout life to survive and grow. In many regions of the CNS, when axons are cut, these stimulatory mechanisms are severely disturbed, leading to neuronal death and loss of regenerative ability. Therefore, to promote regeneration, we will likely need a multi-factorial approach in which inhibition is neutralised and growth is stimulated. One difficult task is successfully applying the same techniques used in animal model experiments to a clinical setting. For example, current concerns centre around the delivery of therapeutic agents in sufficient doses to specific sites within the CNS that has been damaged. One way would be to deliver such agents by subcutaneous pumps that inject antibodies, or other blocking molecules, directly into the cerebrospinal fluid. In addition, reproducibility has been a major problem in experiments designed to promote regeneration following SCI.

As reports of functional regeneration become more numerous, it is important to critically evaluate the mechanisms of the observed functional changes.

Case by case, it will be crucial to consider the inherent plasticity and variability of sparing in each injury model by asking certain key questions, such as the extent to which the sprouting or compensation by spared systems contributes to recovery in each case. The inherent complexity of the adult CNS, as well as its amazing plasticity and adaptiveness make the interpretation of in vivo regenerative experiments in animal models challenging.

Present science has only produced partial cures. Future technological advances will lead to more and better tools, allowing us to address many unanswered questions and to identify the most promising therapeutic approaches for the patient with SCI. The work of the past decade or more has finally produced a variety of treatments that can return partial function to experimental animals. The challenge for the coming decade is to translate these into treatments for human patients. It is not clear if rodent animal models are a sufficient test-bed for determining whether treatments will work in humans. There has been extensive discussion about alternative large animal models in pigs, dogs, cats, and even primates. Some treatments will be tested in these animals before reaching human trials, but some treatments will probably go directly from rodents to humans. Time will tell which is the correct route. Also under discussion is the point at which patients should be treated. Neuroprotective treatments must be given close to the time of injury, and it may be easier to promote axon regeneration shortly after injury. However, treatments that promote plasticity should work in chronically injured patients, of whom there are millions in the world. Until there is success in taking a neuroprotective, transplantation, regenerative, or conduction-enhancing approach from concept to approval, we cannot be sure what to expect along the way.

References

1. Bracken MB et al. (1990) N Engl J Med 322:1405
2. Nobunaga AI et al. (1999) Arch Phys Med Rehabil 80:1372
3. Guttman L (1973) Trans Med Soc Lond 89:228
4. Ramon y Cajal S (1995) Histology of the nervous system of man, vertebrates. Oxford University Press, New York
5. Fawcett JW (1998) Spinal Cord 36:811
6. Gale et al. (1985) Exp Neurol 88:123
7. Noble LJ, Wrathall JR (1985) Exp Neurol 88:135
8. Tator CH, Fehlings MG (1991) J Neurosurg 75:15
9. Behrmann DL et al. (1992) J Neurotrauma 9:197
10. Blight AR (1996) An overview of spinal cord injury models. In: Narayan RK, Povlishock JT, Wilberger JE (eds) Neurotrauma. McGraw-Hill, Toronto, pp 1367–1379
11. Allen A (1914) J Nerv Ment Dis 41:141
12. Ducker TB et al. (1971) J Neurosurg 35:700
13. Dohrmann GJ et al. (1976) Surg Neurol 6:263
14. Blight AR (1983) Neuroscience 10:521
15. Griffiths, McCulloch (1983) J Neurol Sci 58:335
16. Banik NL et al. (1987) Neurochem Pathol 7:57

17. Wallace MC et al. (1987) Surg Neurol 27:209
18. Osterholm JL (1978) The pathophysiology of spinal cord trauma. Thomas, Springfield, IL
19. Guizar-Sahagun et al. (1994) Surg Neurol 41:241
20. Katkulus BA (1984) Cent Nerv Syst Trauma 1:117
21. Bunge RP et al. (1993) Adv Neurol 59:75
22. Tator CH, Rowed DW (1979) Can Med Assoc J 121:1453
23. Choi DW (1992) J Neurobiol 23:1261
24. McIntosh IG (1993) J Neurotrauma 10:215
25. Braken MB et al. (1990) N Engl J Med 322:1405
26. Braken MB et al. (1992) J Neurosurg 76:23
27. Braken MB et al. (1997) JAMA 277:1597
28. Hurlbert RJ (2000) J Neurosurg 93:1
29. Pointillart V et al. (2000) Spinal Cord 38:71
30. Geisler FH, Dorsey FC, Coleman WP (1991) N Engl J Med 324:1829
31. Caquil-Caubere I, Oxhamre C, Kamanka JM, Barbanel G (1999) J Neurosci Res 56:160
32. Pencalet P, Ohanna F, Poulat P, Kamenka JM, Privat A (1993) J Neurosurg 78:603
33. Gaviria M, Privat A, d'Arbigny P, Kamenka JM, Haton H, Ohanna F (2000) Brain Res 874:200
34. Rosenberg LJ, Teng YD, Wrathall JR (1999) J Neurosci 19:464
35. Bethea et al. (1999) J Neurotrauma 16:851
36. Hatten ME et al. (1991) Glia 16:779
37. David S, Aguayo AJ (1981) Science 214:931
38. Fawcett JW, Asher RA (1999) Brain Res Bull 49:377
39. Ridet JL et al.. (1997) Trends Neurosci 20:570
40. Fitch MT, Silver J (1999) Beyond the glial scar. Cellular, molecular mechanisms by which glial cells contribute to CNS regenerative failure. In: Tuszynski MH, Kordower JH (eds) CNS regeneration. Basic science, clinical advances. Academic, New York, pp 55–88
41. Caroni P, Schwab ME (1988) Neuron 1:85
42. Kalderon N, Fuks Z (1996) Proc Natl Acad Sci USA 93:11179
43. Kalderon N, Fuks Z(1996) Proc Natl Acad Sci USA 93:11185
44. Ridet JL, Privat A (2000) Exp Neurol 161:1
45. Rhodes KE, Moon LD, Fawcett JW (2003) Neuroscience 120:41
46. Ridet I, Privat A (2000) Trends Neurosci 23:58
47. Logan A et al. (1994) Eur J Neurosci 6:355
48. Moon LD, Fawcett JW (2001) Eur J Neurosci 14:1667
49. Balasingam V, Yong VW (1996) J Neurosci 16:2945
50. Brewer KL et al. (1999) Exp Neurol 159:484
51. Chen MS, Huber AB, van der Haar ME, Frank M, Schnell L, Spillmann AA, Christ F, Schwab ME (2000) Nature 403:434
52. Schnell L, Schwab ME (1990) Nature 343:269
53. Bregman BS, Kunkel-Bagden E, Schnell L, Dai HN, Gao D, Schwab ME (1995) Nature 378:498
54. Huang DW, McKerracher L, Braun PE, David S (1999) Neuron 24:639
55. Dyer JK, Bourque JA, Steeves JD (1998) Exp Neurol 154:12
56. Rapalino O, Lavarov-Spiegler O, Agranov E, Velan GJ, Yoles E, Fraidakis M, Solomon A, Gepstein R, Katz A, Belkin M, Hadani M, Schwartz (1998) Nat Med 4:814
57. Simonen M, Pedersen V, Weinmann O, Schnell L, Buss A, Ledermann B, Christ F, Sansig G, Van der Putten H, Schwab ME (2003) Neuron 38:201
58. GrandPre T, Li S, Strittmatter SM (2002) Nature 417:547

59. Fournier AR, GrandPre T, Strittmatter SM (2001) Nature 409:341
60. Wang KC, Koprivica V, Kim JA, Sivasankaran R, Guo Y, Neve RL, He Z. (2002) Nature 417:941
61. Liu BP, Fournier AR, GrandPre T, Strittmatter SM (2002) Science 297:1190
62. Zheng B, Ho C, Li S, Keirstead H, Steward O, Tessier-Lavigne M (2003) Neuron 38:213
63. Kim JE, Li S, GrandPre T, Qui D, Strittmatter SM (2003) Neuron 38:187
64. McKeon RJ, Schreiber RC, Rudge JS, Silver J (1991) J Neurosci 11:3398
65. Lemons ML et al. (1999) Exp Neurol 160:51
66. Yick et al. (2000) Neuroreport 11:1063
67. Bradbury EJ, Moon LD, Popat RJ, King VR, Bennett GS, Patel PN, Fawcett JW, McMahon SB (2002) Nature 416:636
68. Moon LD, Asher RA, Rhodes KE, Fawcett JW (2001) Nat Neurosci 4:465
69. Pizzorusso T, Medini P, Berardi N, Chierzi S, Fawcett JW, Maffei L (2002) Science 298:1248
70. Stichel CC, Muller HW (1998) Prog Neurobiol 56:119
71. Stichel CC et al. (1999a) Eur J Neurosci 11:632
72. Stichel CC et al. (1999b) Neuroscience 93:321
73. Weidner et al. (1999) Exp Neurol 160:40
74. Stichel CC, Lausberg F, Hermanns S, Muller HW (1999) Restor Neurol Neurosci 15:1
75. Shearer MC, Niclou SP, Brown D, Asher RA, Holtmaat AJ, Levine JM, Verhaagen J, Fawcett JW (2003) Mol Cell Neurosci 24:913
76. Pasterkamp RJ, Giger RJ, Ruitenberg MJ, Holtmaat AJ, de Wit J, De Winter F, Verhaagen J (1999) Mol Cell Neurosci 13:143
77. Neumann S, Woolf CJ (1999) Neuron 23:83
78. Qui J, Cai D, Dai H, McAtee M, Hoffman PN, Bregman BS, Filbin MT (2002) Neuron 34:895
79. Cai D, Shen Y, DeBellard M, Tang S, Filnin MT (1999) Neuron 22:89
80. Wang KC, Kim JA, Sivasankaran R, Segal R, He Z (2002) Nature 420:74
81. Fournier AE, Takizawa BT, Strittmater SM (2003) J Neurosci 23:1416
82. Yamashita T, Tucker KL, Barde YA (1999) Neuron 24:585
83. Dergham P, Ellezam B, Essagian C, Avedissian H, Lubell WD, McKerracher L (2002) 22:6570
84. Lehmann M, Fournier A, Selles-Navarro I, Dergham P, Sebok A, Leclerc N, Tigyi G, McKerracher L (1999) J Neurosci 19:7537
85. Ellezam B, Dubreuil C, Winton M, Loy L, Dergham P, Selles-Navarro I, McKerracher L (2002) Prog Brain Res 137:371
86. Privat A, Pencalet P, Gimenez-Ribotta M, Mersel M, Rajaofetra NU (1993) Agressologie 2:64
87. Fawcett JW (1992) Trends Neurosci 15:5–8
88. Li D, Field PM, Raisman G (1995) Eur J Neurosci 7:1164
89. Condic ML, Snow DM, Letorneau PC (1999) J Neurosci 19:10036
90. Lahrtz F, Horstkorte R, Cremer H, Schachner M, Montag D (1997) J Neurosci Res 50:62
91. Tetzlaff W, Kobayashi NR, Giehl KM, Tsui BJ, Cassar SL, Bedard AM (1994) Prog Brain Res 103:271
92. Doster SK, Lozano AM, Aguayo AJ, Willard MB (1991) Neuron 6:635
93. Takei Y, Teng J, Harada A, Hirokawa N (2000) J Cell Biol 150:989
94. Fawcett JW, Mathews G, Housden E, Goedert M, Matus A (1994) Neuroscience 61:789
95. Fernandes KJL, Fan DP, Tsui BJ, Cassar SL, Tetzlaff W (1999) J Comp Neurol 414:496
96. Bregman BS, McAtee M, Dai HN, Kuhn PL (1997) Exp Neurol 148:475
97. Kim D, Schallert T, Liu Y, Browarak T, Nayeri N, Tessler A, Fischer I, Murray M (2001) Neurorehabil Neural Repair 15:141

98. Kobayashi NR, Fan DP, Giehl KM, Bedard AM, Wiegand SJ, Tetzlaff W (1997) J Neurosci 15:9583
99. Rossi F, Buffo A, Strata P (2001) Restor Neurol Neurosci 19:85
100. Fouad K, Pedersen V, Schwab ME, Brosamle C (2001) Curr Biol 11:1766
101. Raineteau O, Schwab ME (2001) Nat Rev Neurosci 2:263
102. Kennard MA (1936) Am J Physiol 115:138
103. Thallmair M, Metz GA, Z'Graggen WJ, Raineteau O, Kartje GL, Schwab ME (1998) Nat Neurosci 1:124
104. Raineteau O, Fouad K, Noth P, Thallmair M, Schwab ME (2001) Proc Natl Acad Sci USA 98:6929
105. Borgens RB, Toombs JP, Breur G, Widmer WR, Waters D, Harbath AM, March P, Adams LG (1999) J Neurotrauma 16:639
106. Bunge MB (2001) Neuroscientist 7:325
107. Ramon-Cueto A, Cordero MI, Santos-Benito FF, Avila J (2000) Neuron 25:425
108. Li Y, Field PM, Raisman G (1997) Science 277:2000
109. McDonald JW, Howard MJ (2002) Prog Brain Res 157:299
110. Geller HM, Fawcett JW (2002) Exp Neurol 174:125
111. Tobias CA, Dhoot NO, Wheatley MA, Tessler A, Murray M, Fischer I (2001) J Neurotrauma 18:287
112. Gautier SE, Oudega M, Fragoso M, Chapon P, Plant GW, Bunge MB, Parel JM (1998) J Biomed Mater Res 42:642
113. McDonald JW, Sadowsky C (2002) Lancet 359:417
114. Gage FH (2000) Science 287:1433
115. Crowe MJ, Bresnahan JC, Shuman SL, Masters JN, Beattie MS (1997) Nat Med 3:73
116. Guest JD, Rao A, Olson L, Bunge MB, Bunge RP (1997) Exp Neurol 148:502
117. Shi R, Blight AR (1997) Neuroscience 77:553
118. Potter PJ, Hayes KC, Segal JL, Hsieh JT, Brunnemann SR, Delaney GA, Tierney DS, Mason D (1998) J Neurotrauma 15:837
119. Wolfe DL, Hayes KC, Hsieh JT, Potter PJ (2001) J Neurotrauma 18:757
120. Liu S, Qu Y, Stewart TJ, Howard MJ, Chakrabortty S, Holekamp TF, McDonald JW (2000) Proc Natl Acad Sci USA 97:6126
121. Sasaki M, Honmou O, Akiyama Y, Uede T, Hashi K, Kocsis JD (2001) Glia 35:26
122. Franklin RJ (2002) Brain Res Bull 57:827
123. Ramon-Cueto A, Plant GW, Avila J, Bunge MB (1998) J Neurosci 18:3803
124. Blight AR (1983) Neurosciences 10:521

Received: February 2004

Adv Biochem Engin/Biotechnol (2005) 94: 67–89
DOI 10.1007/b100000

Peripheral Nerve Regeneration

Mei Zhang · Ioannis V. Yannas (✉)

Department of Mechanical Engineering and Division of Biological Engineering, Massachusetts Institute of Technology, 77 Massachusetts Avenue, Cambridge, MA 02139-4307, USA
meizhang@mit.edu, yannas@mit.edu

Abstract The nerve chamber model has dominated the experimental study of peripheral nerve (PN) regeneration with animal models as well as in several clinical applications, such as the treatment of paralysis of limbs following severe trauma. The two stumps resulting from nerve transection are inserted inside a tubular chamber made from one of several materials, occasionally filled with various substances, and the quality of the reconnected nerve is assayed. Recent use of methods for data reduction has led to generation of a large normalized database from independent investigations. Methods for data normalization (reduction) are based on systematic use of the critical axon elongation, L_c, the gap length between the transected stumps at which the frequency of reconnection is just 50% for a given configuration. Four theories are compared for their ability to explain the normalized data. Although the neurotrophic and contact guidance theories explain some of the data, combined use of the more recent microtube theory and pressure cuff theory appears capable of explaining a much larger data set. PN regeneration appears to be upregulated by chamber configurations that facilitate formation of basement membrane microtubes about 10–20 µm in diameter, comprising linear columns of Schwann cells surrounded by basement membrane, into which axons elongate and eventually become myelinated. Regeneration is downregulated by experimental configurations that permit formation of a contractile cell (myofibroblast) capsule around the regenerating nerve that appears to restrict growth of a nerve trunk by application of circumferential mechanical forces. These two processes work competitively to regulate nerve regeneration in the chamber model.

Keywords Peripheral nerve · Nerve chamber · Nerve regeneration · Microtubes · Myofibroblast capsule

Abbreviations

PN	Peripheral nerve
PNS	Peripheral nervous system
BM	Basement membrane
L_c	Critical axon elongation
ΔL	Length shift

1 Introduction

Approximately 200,000 patients are treated each year in the United States for peripheral nerve (PN) injuries that require surgical intervention [1]. These injuries result from accidental trauma, disease, or surgical procedures that require transection of PNs to gain access to the surgical site. An example of surgical trauma is the procedure used for oncological surgery to remove tumors from the cranial base, in which the facial nerve is severed to provide access to this region [2].

Regardless of the cause of severe PN injury, the untreated nerve is partially or totally paralyzed if left untreated. Two methods are commonly employed clinically: direct suturing of the cut ends (nerve stumps) resulting from severe injury or autografting of the gap between the stumps. Direct suturing techniques are used when the two nerve stumps can be directly apposed without significant tension forming at the suture line. This happens when the gap left between the two stumps is very small, e.g., in the range 1–5 mm for the rat model [1]. However, when this type of treatment was used clinically in the median nerve, only 25% of patients recovered full motor function and 3% full sensory function [3]. Autografting is the current clinical treatment for severed PNs where a significant gap exists between the nerve ends, preventing direct suturing. This procedure requires that an additional site be traumatized to remove an intact sensory nerve (e.g., sural nerve), which serves as the source of autograft tissue. Following autografting of the median nerve, only 20% of patients recovered full motor function and no patient recovered full sensory function [3].

Since as early as 1880, experimental investigators and clinicians alike have bridged the stumps of a transected PN by inserting them in tubes (nerve chambers) fabricated from a large variety of materials. Over 180 publications describing various nerve chamber procedures have been cited in early reviews [4]. Several clinical investigators have employed a nerve chamber in an attempt to bring about functional recovery in humans following severe nerve injury

[1, 5–8]. However, even though reconnection of two transected stumps inside a nerve chamber may occur, the regenerated nerve may not connect accurately to its target organ.

During the past 20 years, following the fundamental studies of Lundborg and coworkers with nerve chambers [9–12], an animal model based on a relatively simple protocol has become standardized. In this model, the gap between the two stumps inside the chamber is arbitrarily adjusted by the investigator to a certain length that appears to provide favorable conditions for testing the experimental configuration. Investigators have also used a number of animal models, based on selection of various species and occasionally different nerves in each species. A large variety of assays for quality of regeneration has also been employed by independent laboratories. The search for devices that greatly upregulate the quality of nerve regeneration could be speeded up significantly if investigators could use standardized assays that help them compare their results against those of independent investigators. However, the investigators' widespread use of a variety of gap lengths, animal species and anatomical sites, as well as assays of quality of regeneration, has generated a large store of data that cannot be compared directly. In the absence of methods for direct analysis of data it is not possible to derive useful empirical rules that summarize the data and provide a platform for theoretical developments.

A specific objective of the authors of this chapter is to describe such a methodology for data analysis and to develop theories that explain the data. The new analytical methodology is used to compare semiquantitatively several chamber devices that induce regeneration, and leads to development of empirical rules. These rules predict which experimental configurations provide high quality of nerve regeneration and which fail in that respect. The empirical conclusions are then used to develop new mechanistic theories for regeneration that are consistent with the independent data. Since the conclusions arrived at appear to extend beyond the rat and the mouse, the commonly used laboratory animals, they are considered to be applicable to humans, specifically those who have been treated for paralysis resulting from traumatic nerve injury.

2
Critical Axon Elongation, L_c: a Normalized Measure of Regenerative Activity

As mentioned in the previous section, the large number of assays employed by independent investigators poses a formidable obstacle to the direct comparison of the efficacy of different chamber configurations. Although each of these outcome assays conveys valuable information about the status of the regenerated nerve, the lack of one or two standardized, widely used assays usually defeats direct comparison between chambers that have been studied by different investigators.

A dimensionless measure of synthesis of a nerve trunk with myelinated axons along a gap is quite often used: the frequency of reinnervation across a gap inside a nerve chamber, reported as percent of nerves (*%N*) – or animals – fitted with a nerve chamber that was bridged by myelinated axons. This assay has been employed in several studies with the sciatic nerve of the rat [9, 13–24] as well as that of the mouse [25–29]. Unfortunately, investigators have typically reported data on *%N* at only a single gap length. Since, as described in detail below, *%N* drops abruptly at a critical gap length, it is not possible to compare *%N* values among different investigators without first correcting the data for differences in gap length. Furthermore, since a given gap length in a rat nerve represents a distinctly different experimental condition than does the same gap length in the mouse, a problem of data reduction with respect to animal species must also be dealt with. The data reduction procedure described below corrects both for differences in gap length and differences in animal species among investigators.

The normalization procedure is based on early data showing that a relatively small increase in the gap length bridged by a silicone tube is followed by a sharp drop in *%N* [9]. The available data have been used to construct a "characteristic curve" for the silicone tube configuration (Fig. 1). In this S-shaped curve, the *critical axon elongation*, L_c, can be measured at the inflexion point; it is the gap length beyond which *%N* drops below 50%. The data in Fig. 1 show that L_c=9.7±1.8 mm for the rat sciatic nerve bridged with an unfilled silicone tube (in some cases the silicone tube was prefilled with a buffer that had no incremental regenerative activity compared to the unfilled tube). Given the widespread use of the unfilled silicone tube in studies of regeneration in the

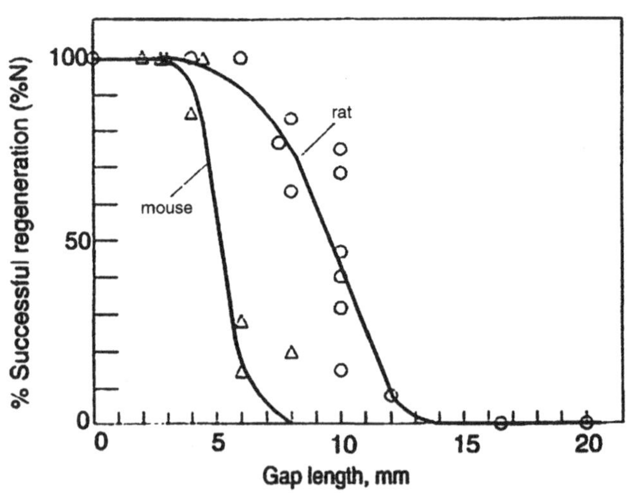

Fig. 1 The characteristic curve for an unfilled silicone chamber used to bridge gaps of variable length in the sciatic nerve of the rat and mouse. The critical axon elongation, L_c, is defined as the gap length at which successful regeneration, *%N*=50, signifying an even chance of reconnection across the gap with a conducting nerve trunk. L_c is 9.7±1.8 mm for the rat and 5.4±1.0 mm for the mouse

peripheral nervous system (PNS), we will use the characteristic curve for this tube to define the outgrowth of axons in a "standard" device.

By comparing their respective characteristic curves against the silicone standard, it will now be possible to analyze the relative performances of other nerve chambers. This simple rationale is based on the expectation that the length of the gap across which a device can induce outgrowth of axons is a direct measure of the regenerative activity of the device. For a given experimental configuration the incremental regenerative activity compared to the standard is defined in terms of the incremental outgrowth of axons across the gap inside the nerve chamber, i.e., as the increase in L_c relative to the standard. The validity of this simple assumption is discussed below by considering its predictive power for a number of different chamber configurations. For example, a chamber that is made of collagen induces outgrowth of axons along a much longer gap length (i.e., has much higher regenerative activity) than does a silicone chamber.

Accordingly, the difference between the values of L_c of an "unknown" or test device and the silicone chamber, the *length shift*, ΔL, will be used as a measure of the regenerative activity of a chamber. A large number of chambers can now be compared directly provided that all are compared against the same standard (silicone chamber). Generalizing along the same direction, we see that the additional gap length (increased regenerative activity) conferred by a specific parameter X of a device, e.g., the chemical composition or enhanced cell permeability of a chamber, can be estimated by comparing L_c for the chamber in the presence and absence of the parameter under study [30].

Provided that the relationship between $\%N$ and the gap length, L, is known for a new chamber, it is possible to construct the characteristic curve for the device and thereby estimate ΔL by comparing it with the standard. Unfortunately, for most experimental configurations such data have not been reported. In the typical investigation reported in the literature, only a single data point, a value of $\%N$ at the investigator's arbitrarily chosen gap length, is reported. To account for the lack of data it is hypothesized here that the characteristic curves of various devices are identically S-shaped and differ from the curve for the silicone standard only by a simple horizontal shift along the gap length axis (typically to the right, indicative of the lower regenerative activity of the silicone standard). This assumption amounts to considering that there are basic similarities in the process of regeneration inside any chamber; specific similarities can be identified below in the discussion of the nature of L_c. On the basis of this assumption it is possible to construct the characteristic curve for a test chamber using the very limited data used by most investigators when reporting the efficacy of their devices, often a single experimental value of $\%N$ at a given gap length. The advantages and weaknesses of this extrapolation procedure have been discussed in detail elsewhere [30].

The procedure described above has been used to construct Table 1, which lists some 26 entries, based on data from independent investigators of nerve regeneration who inserted the stumps of a transected nerve inside a chamber,

Table 1 Regenerative activity of several chamber configurations. Data is from the rat sciatic nerve. Regenerative activity of a configuration is expressed in terms of the critical axon elongation, L_c

Experimental variable X	L_c in presence of X, mm	L_c in absence of X, mm	Shift length, ΔL, mm[a]	References
A. Effects of tubulation, insertion of distal stump and ligation of distal chamber end				
Collagen chamber vs no tubulation	≥13.4	≤6.0	>7.4	Chamberlain et al. 2000 [45]
Distal stump inserted vs open-ended chamber	11.7	≤6.0	>5.7	Williams et al. 1984 [13]
Distal stump inserted vs ligated distal end	11.7	≤6.0	>5.7	Williams et al. 1984 [13]
B. Chamber wall composition				
Silicone chamber (standard)	9.7	(9.7)[b]	0	Data in Fig. 1
EVA copolymer[c] vs silicone standard	≤11.0	(9.7)[b]	≤1.3	Aebischer et al. 1989 [65]
PLA, plasticized[d] vs silicone standard	≥13.4	(9.7)[b]	≥3.7	Seckel et al. 1984 [15]
LA/ε-CPL[e] vs silicone standard	≥13.4	(9.7)[b]	≥3.7	den Dunnen et al. 1993 [22]
Collagen chamber vs silicone chamber	≥13.4	8.0	≥5.4	Chamberlain et al. 1998 [24]
C. Chamber wall permeability				
Cell-permeability vs impermeability	≥19.4	13.4	≥6.0	Jenq and Coggeshall, 1987 [69]
Cell-permeability vs protein-permeability	≥11.4	13.4	≥–2	Jenq et al. 1987 [70]
Protein-permeability vs impermeability	7.5	8.9	–1.4	Jenq and Coggeshall, 1985 [48]; Jenq et al. 1987 [70]

Table 1 (continued)

D. Schwann cell suspensions				
Schwann cell suspension vs PBS	21.4	≤14.0	≥7.4	Ansselin et al. 1997 [71]
E. Chamber filling: solutions of proteins				
bFGF vs no factor	14.3	≤11.0	>3.3	Aebischer et al. 1989 [65]
NGF vs cyt C^f	11.1	10.4	0.7	Hollowell et al. 1990 [74]
aFGF vs no factor	15.5	≤11.0	≥4.5	Walter et al. 1993 [66]
F. Chamber filling: gels based on ECM components				
Fibronectin vs cyt C^f	19.1	18.6	0.5	Bailey et al. 1993 [21]
Laminin vs cyt C^f	18.6	18.6	0	Bailey et al. 1993 [21]
G. Chamber filling: insoluble substrates				
Collagen-GAG matrix vs no matrix	16.1	≤11.0	≥5.1	Yannas et al. 1985 [76], 1987 [38, 77]
Oriented fibrin matrix across gap vs oriented matrix only adjacent to each stump	15.5	≤11.0	≥4.5	Williams et al. 1987 [14]
Early-forming vs late-forming fibrin matrix	15.5	12.5	3.0	Williams et al. 1987 [14]
Axially vs randomly oriented fibrin	11.4	≤6.7	≥4.7	Williams, 1987 [18]
Rapidly degrading CG matrix (NRT^g) vs no CG matrix	≥13.4	8.5	≥4.9	Yannas et al. 1988 [75]; Chang et al. 1990 [20]; Chang and Yannas, 1992 [39]
Axial vs radial orientation of pore channels in NRT^g	≥13.4	10.0	≥3.4	Chang et al. 1990 [20]; Chang and Yannas, 1992 [39]

Table 1 (continued)

Experimental variable X	L_c in presence of X, mm	L_c in absence of X, mm	Shift length, ΔL, mm[a]	References
Laminin-coated collagen sponge vs no laminin coating	11.7	≤6.0	>5.7	Ohbayashi et al. 1996 [78]
Polyamide filaments[h] vs no filaments	18.4	≤11.0	≥7.4	Lundborg et al. 1997 [41]
NRT[g] in collagen chamber vs silicone chamber[i]	>25	7.7	>17.3	Spilker, 2000 [79]

[a] Regenerative activity increases with the shift length, ΔL, the difference between the L_c value for the test configuration and the internal control (or the unfilled silicone chamber standard). Standard: silicone chamber, L_c=9.7.

[b] Parentheses indicate use of the value for the silicone standard (used in the absence of internal control data).

[c] *EVA* Ethylene-vinyl acetate copolymer (Aebischer et al. 1989 [65]).

[d] *PLA* Poly(lactic acid) (Seckel et al. 1984 [15]).

[e] *LA/ε-CPL* Copolymer of lactic acid and ε-caprolactone (den Dunnen et al.1993 [22, 80]).

[f] *Cyt C* Cytochrome C.

[g] *NRT* Nerve regeneration template, a graft copolymer of type I collagen and chondroitin 6-sulfate, differing from dermis regeneration template by a higher degradation rate and an axial (rather than random) orientation of pore channel axes along the nerve axis.

[h] Eight polyamide filaments, each 250 μm in diameter, placed inside the silicone chamber.

[i] Obtained by implanting the experimental chamber in the cross-anastomosis (CA) surgical procedure (Lundborg et al. 1982 [9]) in which both contralateral sciatic nerves of the rat are transected; the right sciatic nerve is transected proximally at the sciatic notch allowing the right distal segment to be placed near the left proximal stump inside the opposite ends of the experimental chamber. The CA procedure allows study of gap lengths over 20 mm.

either unfilled or filled with test substances or cell suspensions. Most of the data were obtained in studies that were specifically designed to study the effect of a single device parameter, e.g., the chemical composition of the tube wall or the addition of a cytokine solution inside the gap. The entries allow direct comparison of incremental regenerative activity (incremental outgrowth of axons across nerve chamber) between two chambers that differed only by a single device parameter *X*. Such direct comparison facilitates an assessment of the isolated impact of a specific experimental variable on the regenerative activity of the chamber.

A limitation of this approach stems from the same feature that makes it valuable: the abruptness with which %*N* drops off with gap length in the vicinity of L_c (Fig. 1). Due to this property of PNS regeneration in the chamber model, it is possible to observe values of %*N* that lie between 0 and 100 only within a short interval of the gap length. Values of %*N* observed outside this narrow range of the gap length can only be reported as upper or lower bounds (Table 1). Although still quite useful, entries in Table 1 that are marked by inequality symbols are obviously less informative than would be values of L_c and ΔL obtained with experiments that were performed inside the critical gap range.

A useful byproduct of the normalization procedure based on the use of a common assay and use of L_c is the realization that data obtained with the mouse sciatic nerve fitted with a nerve chamber can be treated in similar fashion as data obtained with the rat sciatic nerve. The result is shown in Fig. 1, an S-shaped curve compiled from independent data for unfilled silicone chambers that have been used to induce regeneration in the rat and the mouse. The critical axon elongation for the mouse sciatic nerve, L_c=5.4±1.0 mm, is significantly lower than that for the rat sciatic nerve. This finding immediately questions the practice of comparing the effectiveness of nerve chambers that were studied with a variety of animal species by simply citing the absolute gap length along which regeneration had been observed. Clearly, the absolute gap length is not meaningful unless it has been normalized by taking into account the species with which it was observed.

A process of normalization of the gap length for each species is obtained simply by division of the gap length, *L*, at which an observation of %*N* was made for a given species, by the critical axon elongation, L_c, for the same species, to yield the reduced gap length, L/L_c (Fig. 2). Superposition of rat and mouse data is obtained following this process of data reduction. The observed superposition of data from two species following a simple reduction procedure related to animal size suggests that the S-shaped relation between %*N* and gap length may extend to other species as well. Such an extension could be used in the future to compare the outcomes of two studies in animals that vary significantly in size. For example, it might be possible to compare the severity of a study conducted in the rat at a 20-mm gap length and that conducted in a monkey at a 50-mm gap. Such comparisons do not appear possible today on the basis of available data where such data reduction is not employed.

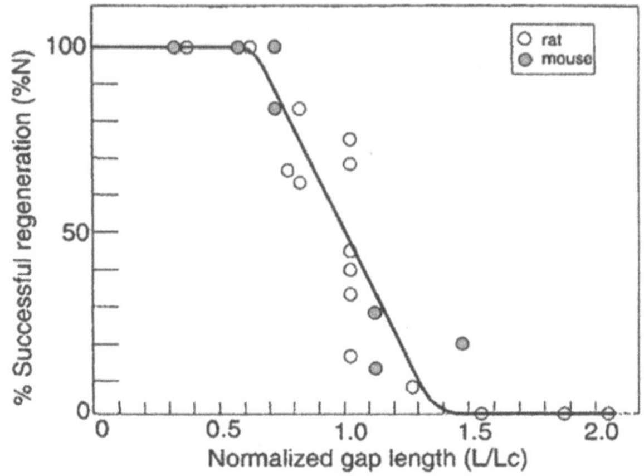

Fig. 2 Superposition of rat (*open circles*) and mouse data (*filled circles*) from Fig. 1 can be obtained by plotting *%N* data against the reduced gap length, L/L_c, the ratio of the gap length at which an observation of *%N* was made in a species divided by the critical axon elongation, L_c, for that species

3
Experimental Configurations that have Significantly Increased the Critical Axon Elongation

Over 20 different chamber fillings and over 20 varieties of chamber type have been studied with the nerve chamber model in the rat and mouse over the past twenty-odd years. An analysis of the efficacy of the various devices used can now be made for devices in which the data include at least a single report of *%N* at a given gap length.

Table 1 is a compilation of data obtained with the rat sciatic nerve model. Mouse data are not shown here; in general, mouse data follow rat data very closely [30]. We recall that ΔL for the 10-mm gap bridged by the silicone tube is, by definition, zero. Positive and negative values of ΔL signify upregulation and downregulation, respectively, of regenerative activity over that for the internal control occasionally employed by an investigator. In the absence of an internal control, the test device was compared to a control consisting of the unfilled silicone tube standard. Given the value for the experimental uncertainty, we arbitrarily consider ΔL values in the range +2 to +4 mm for the rat sciatic nerve model to be "significant" (corresponding to a high value of the regenerative activity); values above +4 mm rat are considered to be "very significant" (corresponding to a very high value of the regenerative activity). Negative values suggest downregulation of regenerative processes relative to the silicone standard. A more detailed assignment of regenerative activity is not warranted by the semiquantitative data.

Some of the conclusions that emerge from the entries in Table 1 agree very well with known, independently obtained, observations in the literature of PNS regeneration [4, 30]. For example, simple insertion of the stumps in a nerve chamber, or insertion only of the proximal stump, resulted in very significant upregulation of regenerative activity compared to an experimental configuration in which the stumps were either not inserted in the nerve chamber or the distal stump was left outside (or was ligated), respectively. Degradable chambers based on each of two synthetic polymers, plasticized poly(lactic acid) and a copolymer of lactic acid and ε-caprolactone, as well as a natural polymer, type I collagen, induced significantly greater outgrowth of axons than degradable chambers based on an ethylene–vinyl acetate copolymer or a nondegradable chamber (silicone). Cell-permeability conferred very significant regenerative activity compared to the impermeable tube; however, a protein-permeable tube lacked such activity.

Use of several chamber fillings resulted in significant variation of the regenerative activity (Table 1). Suspensions of Schwann cells showed very significant regenerative activity as did also use of solutions of acidic (aFGF) and basic fibroblast growth factor (bFGF) but, surprisingly, not solutions of nerve growth factor (NGF). Use of extracellular matrix (ECM) macromolecules, such as collagen, laminin and fibronectin, in the form of solutions or gels had no significant activity; furthermore, when gel concentrations exceeded certain levels, negative activity was noted. Several insoluble substrates showed very significant regenerative activity. Very active substrates included highly oriented fibrin fibers and axially oriented polyamide filaments. Other very active substrates included a family of highly porous ECM analogs characterized by specific surface that exceeded a critical level, by pore channels that were axially oriented and by a sufficiently rapid degradation rate.

4 Mechanism of Nerve Regeneration: How to Explain the L_c Data?

Distinctions among alternative hypotheses for the mechanism of regeneration will now be based on the data in Table 1. Recent advances in the study of the mechanism of PNS regeneration suggest that, in addition to the early neurotrophic theory and the contact guidance theory, two other hypothetical mechanisms must also be considered in order to explain the data in Table 1: the "pressure cuff" theory and the "basement membrane microtube" theory.

4.1 Neurotrophic Theory

In the neurotrophic theory [10, 31–33], elongation of axons and nonneuronal supporting cells from the proximal stump are thought to be controlled by diffusion of growth-promoting soluble factors from the distal stump. This theory is supported by a number of experimental observations [13, 34, 35]. The facts that brain-derived neurotrophic factor (BDNF) and ciliary neurotrophic factor

(CNTF) facilitate PN regeneration, and that an inserted distal stump results in an increase in L_c of at least 5.7 mm compared to the open ended chamber or ligated distal stump (Table 1), are consistent with this theory. However, it appears difficult to explain other data by this theory. For example, the sudden drop in *%N* with a relatively small increase in gap length appears to be inconsistent with the theory; a small increase in gap length should incrementally increase the diffusion path from the emitting stump to the receiving stump without a sudden drop in concentration of neurotrophic factors to account for the relative discontinuity in *%N*. Also, orientation of a substrate should have little predictable effect on the diffusion of neurotrophic factors, yet the regeneration activity is significantly upregulated by the use of oriented insoluble substrates. Finally, the significantly higher regenerative activity of cell-permeable versus protein-permeable chambers is not readily explained by this theory.

4.2 Contact Guidance Theory

In the contact guidance theory it has been originally postulated that axon elongation requires guidance by contact with an appropriate substrate [14, 18, 24, 36–41]. Proliferative and migratory activities of Schwann cells and fibroblasts are further considered to be upregulated in the presence of insoluble substrates, which theoretically provide tracks for cells or elongating axons to attach to, thereby guiding the reconnection process. This theory is widely supported by the observations that the presence and structure of insoluble substrates (e.g., longitudinally oriented pore channels in the collagen-glycosanimoglycan tube filling [39] and magnetically aligned fibrin gels [42]) play an important role in influencing the outcome (Table 1). The theory is also consistent with the observation that the presence of antibodies directed against matrix structures that may normally facilitate regeneration, such as an active site on laminin and an integrin receptor for laminin and collagen, suppresses regeneration [43]. On the other hand, it is conceivable that such substrates promote regeneration by binding on their surfaces the growth factors that are believed to facilitate the process [44]. However, it is hard to use this theory to explain, among other observations, the superior regenerative activity of Schwann cell suspensions or of solutions of certain growth factors.

4.3 Pressure Cuff Theory

The pressure cuff theory originally emerged from the observation of a capsule of contractile cells around the nerve trunk regenerated using a silicone chamber but not around the trunk formed in the presence of a collagen chamber [30, 45]. According to this hypothesis, regeneration across a long enough gap is mechanically blocked by circumferential forces exerted on the regenerating nerve by contractile fibroblasts (myofibroblasts). These contractile forces normally induce closure of the wound generated by transecting the nerve trunk. According to this theory, if the myofibroblast capsule is allowed to develop fully it induces contraction of the diameter of the trunk regenerating nerve and may, under certain conditions, eventually block completely elongation across the gap (neuroma formation).

4.3.1 Experimental Observation of Contractile Cell Capsule

In several early studies investigators observed a relatively thick capsule of cells surrounding the nerve that was regenerated inside silicone chambers. These cells were widely thought to be fibroblasts; a positive identification as myofibroblasts had not been made in these early studies [4, 9–12, 38, 44, 46–52].

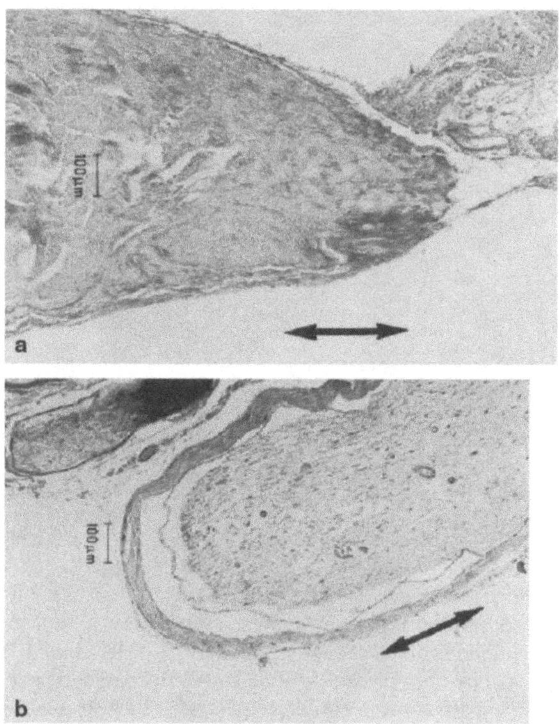

Fig. 3a,b Longitudinal sections of the stumps formed in the absence of a chamber, retrieved after 6 weeks, and stained with Masson's trichrome. The *arrow* indicates the direction of the nerve axis. **a** In the proximal neuroma, a thick, collagen tissue capsule surrounded the nerve tissue and then converged to form a cap at the end of the neuroma. The dense collagenous tissue formed at the end of the neuroma resembled fibrous scar tissue. The tissue capsule around the nerve stump was approximately 20–50 μm thick. **b** In the distal stump, a similar collagen tissue capsule was visible, approximately 50 μm thick, capping off the distal nerve stump. *Scale bars* 100 μm [45]

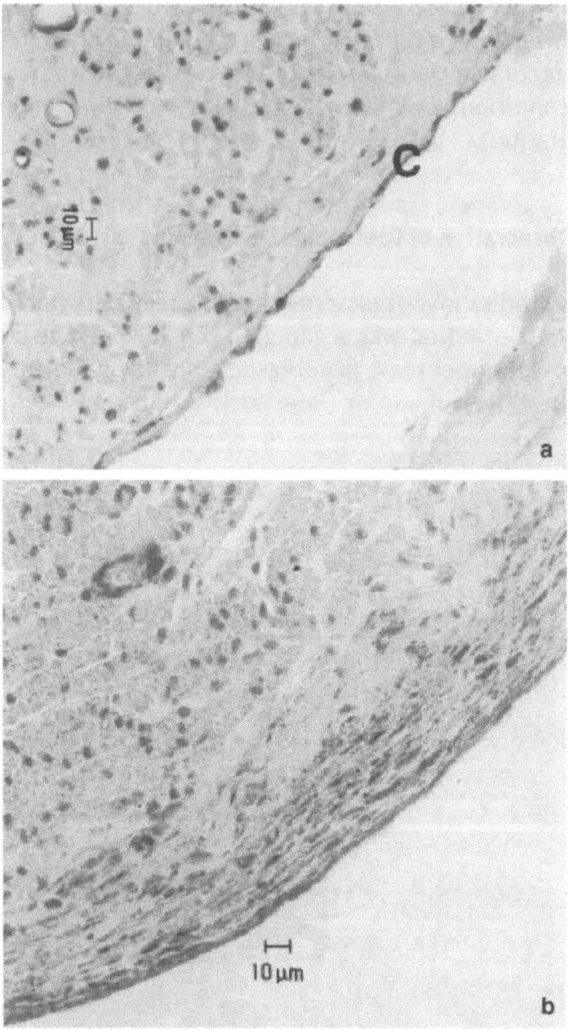

Fig. 4a,b Cross-sectional micrographs of regenerated nerve tissue retrieved after 60 weeks. Both micrographs were stained with the α-smooth muscle actin antibody and viewed at the same magnification. **a** Matrix-filled collagen chamber. The cells of the tissue capsule in the nerve regenerate are stained positively; however the contractile cells (*C*) are not confluent around the entire nerve trunk. In all areas, the contractile cells were no more than one to two cell-layers thick. **b** Matrix-filled silicone chamber. In contrast to the collagen chamber, the nerve regenerate had approximately 15–20 cell layers of myofibroblasts around the entire perimeter of the nerve trunk. *Scale bars* 10 µm [45]

Several authors have independently reported macroscopic contraction in the cross section area of nerve stumps following transection. In some reports contraction was reported as a reduction in cross section area of the distal stump, even to as low as 30–50% of original stump area [36, 53–55]. Although the role of a population of such cells in the healing process was not investigated in earlier studies, identification of isolated contractile cells (myofibroblasts) in a transected nerve was made by ultrastructural observation in two studies [56, 57].

Data reported in the past few years have shown that a thick capsule of contractile cells (myofibroblasts; cells that stain for α-smooth muscle actin), is typically wrapped around a neuroma, i.e., a transected stump that has not reconnected with the stump opposite to it (Fig. 3). Even poorly regenerated nerves that had reconnected were observed to be surrounded by a thick capsule of myofibroblasts. The thickness of the capsule was observed to depend very strongly on the type of chamber used to induce regeneration across a gap. For example, a thick myofibroblast capsule, about 15–20 cell layers in thickness, formed around nerves regenerated inside silicone chambers (Fig. 4) [45, 58]. In contrast, the thickness of the myofibroblast layer around nerves regenerated inside collagen chambers was very thin and quite similar to that which forms around an intact nerve trunk [59]. Furthermore, the quality of regeneration (measured by electrophysiological and morphological properties of regenerated nerves at 60 weeks) of nerves regenerated inside collagen tubes was clearly superior to those regenerated inside silicone tubes [24, 45].

4.3.2 Outcome of Regeneration Theoretically Depends on Contractile Forces Acting Circumferentially on Regenerating Nerve

These observations cannot be explained by the two theories described above. It appears necessary to consider explicitly the healing response of the transected nerve trunk, specifically taking into account the presence of the myofibroblast capsule. The data suggest the presence of an inverse relationship between the quality of regeneration and the thickness of the myofibroblast capsule formed around the regenerated nerve [30].

According to the "pressure cuff" theory, myofibroblasts involved in the normal healing response of the transected nerve stumps induce contraction of the stumps as well as of the regenerating nerve. The latter is viewed primarily as an extension of each wounded stump and its fate is largely controlled by the healing response of the stump. The mechanical forces that are generated in the injured nerve significantly restrict each stump together with its extension consisting of the incipiently regenerating nerve. Similar contractile forces are generated in severe skin wounds, where wound closure has been widely attributed to the activity of myofibroblasts; see review in [30]. In the injured nerve the contractile forces generate circumferential ("hoop") stresses that compress the regenerating nerve radially, theoretically preventing thereby an increase in its diameter and leading to formation of a poorly regenerated nerve or, as in a neuroma, completely preventing any significant growth radially outward [30].

4.3.3 Existence of Critical Gap Length Explained by Pressure Cuff Theory

The pressure cuff theory can be extended to provide an explanation for the existence of a critical axon elongation, measured as L_c. It has been shown that the diameter of the fibrin cable that forms in the gap between the stumps inside the chamber determines the eventual diameter of the regenerating nerve trunk [14, 18, 30, 47]. There is also clear evidence that the cable diameter is limited by the total volume of exudate flowing out of both stumps into the chamber gap [47]. However, the total volume of exudate does not increase with the gap length, as shown by the observation that the cable becomes increasingly thin with increasing gap length [14, 18]. It follows that the diameter of the regenerating nerve should decrease with increasing gap length, as observed [48]; interruption of cable continuity inside the gap should, accordingly, occur when the cable diameter becomes sufficiently thin. A limiting nerve trunk diameter, eventually corresponding to neuroma formation, is therefore, reached when either the circumferential stresses surrounding the cable (and eventually the regenerate) exceed a critical value or when the gap length is sufficiently long; likely, both factors contribute to the observed value of L_c for a given experimental configuration [30].

Data that show how L_c varies with various experimental parameters (Table 1; other morphological data reviewed in [30]), as well as several other observations, can be explained by this theory. As suggested above, the significantly lower value of L_c for the silicone chamber (9.7 mm), compared to that for the collagen chamber (<13.4 mm), is explained as a restriction due to the particularly thick contractile capsule that surrounds the regenerating nerve inside the silicone chamber (but not the collagen chamber) and the resulting high stresses that limit the growth of the regenerating trunk, as observed [24, 45]. The presence of significant circumferential stresses acting symmetrically around the perimeter of the regenerating trunk, but not around the normal (intact) nerve, account for the circular perimeter of the regenerating trunk compared to the frequently elliptical cross section of the intact nerve [48]. The indentations frequently observed in the layers of the connective tissue sheath surrounding the trunk inside the silicone chamber [48] can be explained as buckling of layers that have been loaded above a critical value. The sharp drop in L_c with removal of a stump from the chamber [9, 13] or with ligation of the distal end of the chamber [13] (Table 1) is readily explained as loss of exudate by leakage outside the chamber and blocking of exudate flow from reaching the distal stump, respectively. Omission of the chamber exaggerates these conditions and inevitably leads to neuroma formation, except when the gap length is limited to a very small distance [29]. The increased regenerative activity of cell-permeable chambers relative to impermeable chambers is consistent with the observation [45] that myofibroblasts become trapped inside the porous chamber wall; trapped myofibroblasts are probably migrating through and are conceivably on their way to exiting the chamber, thereby downregulating the contractile stresses exerted on the regenerating nerve and facilitating regeneration. The insignificant effect of protein-permeable chambers compared to impermeable ones [48, 69, 70] may reflect the delivery of a sufficient concentration of proteins, e.g., growth factors, by diffusion from the stumps (neurotrophic theory); additional transfer of proteins through the chamber wall apparently does not have a significant effect. However, other data, e.g., the significant upregulatory effects of solutions of bFGF [65] and aFGF [66] (Table 1), appear to contradict an explanation based on lack of effect of protein-permeability of the chamber.

In addition to its inability to explain the large increase in regenerative activity following the use of solutions of certain growth factors that upregulate Schwann cell activity, the pressure cuff theory also does not explain the very significant increase in L_c following use of Schwann cell suspensions as well as that following several types of chamber fillings (Table 1).

4.4 Basement Membrane Microtube Theory

Observations made by independent investigators suggest strongly that the synthetic events inside the fibrin cable that forms following flow of exudate into the chamber (see above) are dominated by Schwann cell migration along the cable axis. Direct observation has led to the conclusion that, even in experimental protocols in which axons were excluded from the chamber interior, Schwann cells migrated ahead of other cell types, forming long, linear columns [60]. Other data showed that a basement membrane (BM) was formed around Schwann cells that had migrated in the absence of axons [61].

The combined data have led to the theory that Schwann cells migrate along the fibrin cable and synthesize cylindrical "BM microtubes", approximately 10–20 μm in diameter, along the cable axis; in this theory, axons elongate into the preformed microtubes where they also become myelinated [30]. There is substantial evidence that the postulated microtubes resemble, in morphology and regenerative activity, certain organized structures, often referred to as endoneurial tubes (sheaths) [30]; endoneurial tubes are known to form spontaneously in the distal stump, following transection [62].

The microtube theory resembles the contact guidance mechanism, proposed much earlier in order to focus attention on the effect of oriented substrates on axon elongation [36]. Unlike the contact guidance theory, the new theory pays secondary attention to axon elongation, suggesting instead that the critical step is Schwann cell migration along linear paths and their coalescence into columns for eventual axon elongation and myelination.

The microtube theory stated above predicts that an experimental configuration that upregulates synthesis of BM microtubes by Schwann cells facilitates elongation of myelinated axons along the length of the cable, consistent with enhancement of L_c. According to this theory, an experimental configuration that promotes synthesis of BM microtubes by Schwann cells in a very large number (thousands) of independent locations inside the cable cross section, rather than in a few locations, should eventually lead to a nerve trunk characterized by relatively high values of axon density (number of axons per cross-section area of nerve trunk). Predictions from this hypothesis will be compared below with relevant values of L_c and the shift length ΔL in Table 1.

When the chamber is supplied with exogenous Schwann cell suspensions, an exogenous source of migratory Schwann cells supplements the endogenous supply contributed by the stumps; the result of addition of Schwann cells is reflected in the high value of L_c or the shift length, ΔL, as shown in Table 1 [71]. Other studies also show increased quality of regeneration following addition of Schwann cell suspensions [63, 64]. Supply of solutions of bFGF and aFGF, both known as very active mitogens and inducers of migratory behavior of Schwann cells in vitro [67, 68], are expected to upregulate the density of Schwann cells migrating toward the gap center. The observed effect of bFGF [65] and aFGF [66] addition is reflected in substantial increases in L_c (Table 1). On the other hand, the available data show that NGF has much less regenerative activity than either aFGF or bFGF in the chamber configuration [74].

Schwann cells have been shown to start synthesizing NGF after losing axonal contact, following trauma, and to interrupt the synthesis once axonal contact has been reestablished [72, 73].

As shown in Table 1, gels based on fibronectin or laminin (prepared without allowance for orientation of macromolecular constituents prior to implantation) did not have significant regenerative activity [21]. It is hypothesized that gel formation in vivo prevents orientation inside the fibrin cable by trapping either fibrinogen, the monomer of fibrin, or fibrin polymer in a relatively isotropic, semisolid medium; this hypothetical configuration should block formation of linear Schwann cell columns along the nerve axis, leading to a relatively low value of L_c, as observed (Table 1). In contrast, a very high regenerative activity was observed when a highly oriented fibrin matrix was established inside the gap; or when the oriented fibrin matrix was established in 24 h rather than in several days [14, 18]. Furthermore, use of six polyamide filaments (diameter 250 μm), also led to a very high value of regenerative activity [41]. It is speculated that a large variety of types of filament upregulate Schwann cell migration by contact guidance along linear paths, thereby facilitating microtube formation across the gap.

Valuable information about the substrate preferences of Schwann cells can be obtained following use of a homologous series of highly porous ECM analogs in which a structural parameter of interest is changed gradually while the levels of other parameters are kept fixed (Table 1). An ECM analog in which the pores were highly oriented along the major axis led to a very highly significant increase in regenerative activity compared to one in which the pore channels were oriented along a direction perpendicular to this axis (radial orientation) [20, 39]. Likewise, a rapidly degrading ECM analog exhibited a much higher regenerative activity than one with a much slower degradation rate [20, 39, 75]. In another series of ECM analogs the average pore diameter was varied in the range 5–300 μm, corresponding to an increase in specific surface by about 50 times along the direction of decreasing pore size. In this series the quality of regeneration progressively increased as the smaller pore diameters were approached, reaching its highest value near 5 μm [20, 39]. It is conceivable that degradable and highly oriented channels provide, if thin enough, temporary templates in which linear columnary Schwan cells are formed.

Certain studies with ECM analogs were based on the use of electrophysiological rather than morphological assays; accordingly, conclusions from these studies do not appear in Table 1 because L_c data are not available.

In general, the data obtained with well-defined substrates could be explained by the microtube theory that accounts for migration of Schwann cells along linear tracks inside the gap.

5 Conclusion: Two Major Processes Apparently Regulate PN Regeneration

While the formation of a myofibroblast capsule has been theorized to down-regulate regeneration, formation of linear columns of Schwann cells has the opposite effect. The capsule theory explains well many data on the critical axon elongation; the Schwann cell theory provides satisfactory qualitative explanation of data on substrate preferences of regenerating nerves. Since the combined theories appear to explain the great majority of data in Table 1 it is likely that these two parallel processes play a dominant role in PN regeneration. However, the importance to regenerative processes of diffusion of soluble proteins across the gap, described by the neurotrophic theory, is not ruled out by the data. The theoretical mechanisms of regeneration in the tubulation model and the speculative pathways by which chamber parameters affect the quality of regeneration are illustrated in Fig. 5 and Fig. 6, respectively.

Much remains to be understood about the detailed processes that determine the quality of PN regeneration. However, the use of procedures for normalization of data with nerve chambers should increase greatly the ability of investigators to compare data directly and thereby to formulate improved theories of PN regeneration.

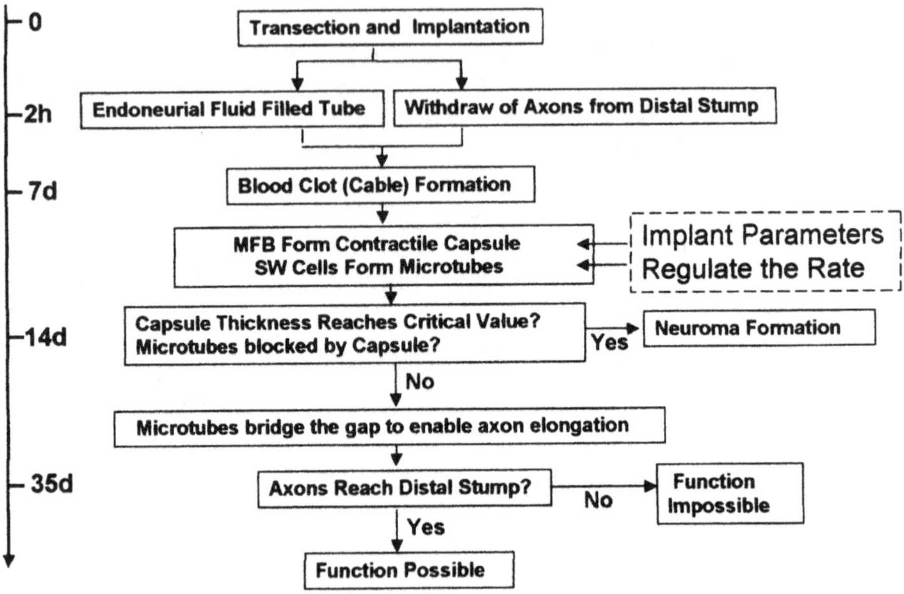

Fig. 5 Hypothesized mechanisms of peripheral nerve regeneration: outcome of regeneration depends on both upregulation by synthesis of microtubes and downregulation by formation of contractile cells capsule. *d* Days

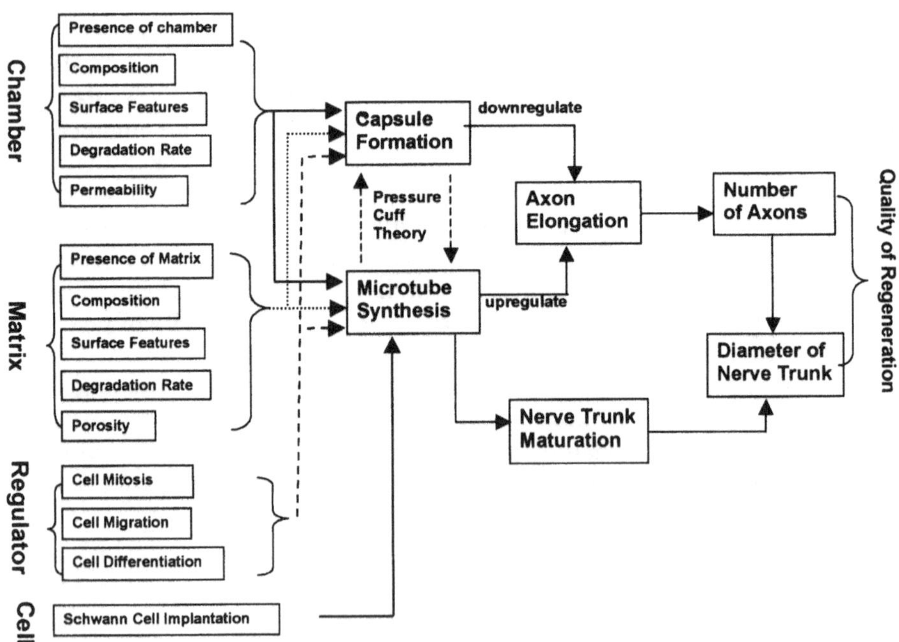

Fig. 6 Effects of chamber parameters on quality of regeneration through regulation of the synthesis of two types of tissue: contractile cell capsule and microtubes

References

1. Madison RM, Archibald SJ, Krarup C (1992) Peripheral nerve injury. In: Cohen IK, Diegelmann RF, Lindblad WJ (eds) Wound healing. Saunders, Philadelphia, p 450
2. Janecka IP, Sen CN, Sekhar LN, Arriaga M (1990) Otolaryngol Head Neck Surg 103:413
3. Mackinnon SE, Dellon AL (1988) Surgery of the peripheral nerve. Thieme, New York
4. Fields RD, Le Beau JM, Longo FM, Ellisman MH (1989) Prog Neurobiol 33:8
5. Lundborg G (1987) Acta Orthop Scand 58:145
6. Weber RA, Breidenbach WC, Brown RE, Jabaley ME, Mass DP (2000) Plast Reconstr Surg 106:1036
7. Dahlin LB, Anagnostaki L, Lundborg G (2001) Scand J Plast Reconstr Surg Hand Surg 35:2
8. Lundborg G, Rosen B, Dahlin L, Holmberg J, Rosen I (2004) J Hand Surg [Br] 29:100
9. Lundborg G, Fahlin LB, Danielsen N, Gelberman RH, Longo FM, Powell HC, Varon S (1982) Exp Neurol 76:361
10. Lundborg G, Dahlin LB, Danielsen N, Johannesson A, Hansson HA, Longo F, Varon S (1982) J Hand Surg 7:580
11. Lundborg G, Gelberman RH, Longo FM, Powell HC, Varon S (1982) J Neuropathol Exp Neurol 41:412
12. Lundborg G, Longo FM, Varon S (1982) Brain Res 232:157
13. Williams LR, Powell HC, Lundborg G, Varon S (1984) Brain Res 293:201
14. Williams LR, Danielsen N, Muller H, Varon S (1987) J Comp Neurol 264:284
15. Seckel BR, Chiu TH, Nyilas E, Sidman RL (1984) Plast Reconstr Surg 74:173
16. Jenq CB, Coggeshall RE (1984) Brain Res 295:91
17. Fields RD, Ellisman MH (1986) Exp Neurol 92:48
18. Williams LR (1987) Neurochem Res 12:851
19. Muller H, Shibib K, Friedrich H, Modrack M (1987) Exp Neurol 95:21
20. Chang AS-P, Yannas IV, Perutz S, Loree H, Sethi RR, Krarup C, Norregaard TV, Zervas NT, Silver J (1990) Electrophysiological study of recovery of peripheral nerves regenerated by a collagen-glycosaminoglycan copolymer matrix. In: Gebelein CG (ed) Progress in biomedical polymers. Plenum, New York, p107
21. Bailey SB, Eichler ME, Villadiego A, Rich KM (1993) J Neurocytol 22:176
22. den Dunnen WFA, van der Lei B, Schakenraad JM, Blaauw EH, Stockros I, Pennings AJ, Robinson PH (1993) Microsurgery 14:508
23. Derby A, Engleman VW, Frierdich GE, Neises G, Rapp SR, Roufa DG (1993) Exp Neurol 119:176
24. Chamberlain LJ, Yannas IV, Hsu H-P, Strichartz G, Spector M (1998) Exp Neurol 154:315
25. Da Silva CF, Madison R, Dikkes P, Chiu T-H, Sidman RL (1985) Brain Res 342:307
26. Aebischer P, Guenard V, Winn SR, Valentini RF, Galletti PM (1988) Brain Res 454:179
27. Santos PM, Winterowd J, Allen GG, Bothwell MA, Rubel EW (1991) Otolaryngol Head Neck Surg 105:12
28. Navarro X, Rodriguez FJ, Labrador RO, Buti M, Ceballos D, Gomez N, Cuadras J, Perego G (1996) J Peripher Nerv Sys 1:53
29. Buti M, Verdu E, Labrador RO, Vilches JJ, Fores J, Navarro X (1996) Exp Neurol 137:26
30. Yannas IV (2001) Tissue and organ regeneration in adults. Springer, New York
31. Cajal RY (1928; reissued 1982) Degeneration and regeneration of the nervous system. Oxford University Press, London
32. Longo FM, Skaper SD, Manthorpe M, Williams LR, Lundborg G, Varon S (1983) Exp Neurol 81:756
33. Longo FM, Manthorpe M, Skaper SD, Lundborg G, Varon S (1983) Brain Res 261:109

34. Ide C (1996) Neurosci Res 25:101
35. Ho PR, Coan GM, Cheng ET, Niell C, Tarn DM, Zhou H, Sierra D, Terris DJ (1998) Arch Otolaryngol Head Neck Surg 124:761
36. Weiss P (1944) J Neurosurg 1:400
37. Weiss P, Taylor AC (1944) J Exp Zool 95:233
38. Yannas IV, Orgill DP, Silver J, Norregaard T, Zervas NT, Schoene WC (1987) Regeneration of sciatic nerve across 15-mm gap by use of a polymeric template. In: Gebelein CG (ed) Advances in biomedical materials. American Chemical Society, Washington DC, p 1
39. Chang AS, Yannas IV (1992) Peripheral nerve regeneration. In: Smith B, Adelman G (eds) Neuroscience year. Birkhäuser, Boston, p 125
40. Whitworth IH, Brown RA, Dore C, Green CJ, Terenghi G (1995) J Hand Surg 20B:429
41. Lundborg G, Dahlin L, Dohi D, Kanje M, Terada N (1997) J Hand Surg (Br Eur) 22B:299
42. Dubey N, Letourneau PC, Tranquillo RT (2001) Biomaterials 22:1065
43. Carbonetto S (1991) Curr Opin Neurobiol 1:407
44. Madison RD, Silva CFD, Dikkes P (1988) Brain Res 447:325
45. Chamberlain LJ, Yannas IV, Hsu H-P, Spector M (2000) J Comp Neurol 417:415
46. Williams LR, Longo FM, Powell HC, Lundborg G, Varon S (1983) J Comp Neurol 218:460
47. Williams LR, Varon S (1985) J Comp Neurol 231:209
48. Jenq C-B, Coggeshall RE (1985) Brain Res 326:27
49. Jenq C-B, Coggeshall RE (1985) Brain Res 345:34
50. Hurtado H, Knoops B, Van den Bousch de Aguilar P (1987) Exp Neurol 97:751
51. Azzam NA, Zalewski AA, Williams LR, Azzam RN (1991) J Comp Neurol 314:807
52. Itoh S, Takakuda K, Samejima H, Ohta T, Shinomiya K and Ichinose S (1999) J Mater Sci Mater Med 10:129
53. Holmes W Young JZ (1942) J Anat (London) 77:63
54. Weiss P Taylor AC (1944) Proc Soc Exp Biol 55:77
55. Sunderland S (1990) Muscle & Nerve 13:771
56. Scaravilli F (1984) J Anat 139:411
57. Badalamente MA, Hurst LC, Eilstein J, McDevitt CA (1985) J Hand Surg 10:49
58. Chamberlain LJ, Yannas IV, Arrizabalaga A, Hsu H-P, Norregaard TV, Spector M (1998) Biomaterials 19:1393
59. Ross MH, Reith EJ (1969) Science 165:604
60. Zhao Q, Dahlin LB, Kanje M, Lundborg G (1992) Brain Res 592:106
61. Ikeda K, Oda Y, Tomita K, Nomura S, Nakanishi I (1989) J Electron Miscrosc (Tokyo) 38:230
62. Fu SY, Gordon T (1997) Mol Neurobiol 14:67
63. Guénard V, Kleitman N, Morrissey TK, Bunge RP, Aebischer P (1992) J Neurosci 12:3310
64. Kim DH, Connoly SE, Kline DG, Voorhies RM, Smith A, Powell M, Yoes T, Daniloff JK (1994) J Neurosurg 80:254
65. Aebischer P, Salessiotis AN, Winn SR (1989) J Neurosci Res 23:282
66. Walter MA, Kurouglu R, Caulfield JB, Vasconez LO, Thompson JA (1993) Lymphokine Cytokine Res 12:135
67. Krikorian D, Manthorpe M, Varon S (1982) Dev Neurosci 5:77
68. Burgess WH, Maciag T (1989) Annu Rev Biochem 58:575
69. Jenq C-B, Coggeshall RE (1987) Brain Res 408:239
70. Jenq C-B, Jenq LL, Coggeshall RE (1987) Exp Neurol 97:662
71. Ansselin A D, Fink T, Davey DF (1997) Neuropathol Appl Neurobiol 23:387
72. Taniuchi M, Clark HB, Johnson EM (1986) Proc Natl Acad Sci USA 83:4094
73. Taniuchi M, Clark HB, Schweitzer HB, Johnson EM (1988) J Neurosci 8:664
74. Hollowell JP, Villadiego A, Rich KM (1990) Exp Neurol 110:45

75. Yannas IV, Chang AS, Krarup C, Sethi R, Norregaard TV, Zervas NT (1988) Soc Neurosci Abstr 14:165.
76. Yannas IV, Orgill DP, Silver J, Norregaard TV, Zervas NT, Schoene WC (1985) Trans Soc Biomater 8:146
77. Yannas IV, Norregaard TV, Silver J, Zervas NT, Kirk JF, Colt MJ (1987) Trans Soc Biomater 10:6
78. Ohbayashi K, Inoue HK, Awaya A, Kobayashi S, Kohga H, Nakamura M, Ohye C (1996) Neurol Med Chir (Tokyo) 36:428
79. Spilker MH (2000) Ph.D. thesis, Massachusetts Institute of Technology
80. den Dunnen WFA, Schakenraad JM, Zondervan GZ, Pennings AJ, van der Lei B, Robinson PH (1993) J Mater Sci Mater Med 4:521

Received: April 2004

Adv Biochem Engin/Biotechnol (2005) 94: 91–123
DOI 10.1007/b100001

Regeneration of Articular Cartilage

B. Kinner[1] · R. M. Capito[2,3] · M. Spector[3] (✉)

[1] Department of Trauma Surgery, Clinics of the University of Regensburg, Regensburg, Germany
drkinner@aol.com

[2] Department of Materials Science and Engineering, Massachusetts Institute of Technology, 77 Massachusetts Avenue Cambridge, MA 02139, USA
rmc_16@mit.edu

[3] Orthopaedic Research Laboratory, Brigham and Women's Hospital, MRB 106, Harvard Medical School, 75 Francis Street, Boston, MA 02115, USA
mspector@rics.bwh.harvard.edu
and Tissue Engineering, VA Boston Healthcare System, Boston Campus, 150 S. Huntingdon Avenue, Boston, MA02130, USA

Abstract Loss of articular cartilage from the ends of bones forming diarthrodial joints can be the source of profound pain and disability, and eventually lead to complete degeneration of the joint, necessitating total joint replacement. Until a few years ago, there seemed little hope of treating such defects. Novel surgical procedures and cell therapies have recently been found, however, to stimulate the formation of reparative tissue resulting in the relief of pain and restoration of function, at least for a limited time period. Moreover, studies of the healing of chondral defects in animal models have revealed that there is some potential for regeneration of this connective tissue. The introduction of certain biomaterial scaffolds along with selected surgical procedures and cell therapies has been demonstrated in animal studies to facilitate the cartilage reparative process and now offers the promise of extending the longevity of clinical treatments of cartilage defects. Collectively these findings provide the basis for the rational development of approaches for the more complete regeneration of articular cartilage, and demonstrate that meaningful clinical outcomes can be achieved even if complete regeneration is not achieved.

Keywords Articular cartilage · Collagen · Chondrocyte · Joints · Microfracture

Abbreviations and Symbols

3-D	Three-dimensional
BMP	Bone morphogenetic protein
ACI	Autologous chondrocyte implantation
CAC	Cultured autologous chondrocyte
DNA	Deoxyribonucleic acid
ECM	Extracellular matrix
FGF	Fibroblast growth factor
GAG	Glycosaminoglycan
IGF	Insulin-like growth factor
PGA	Polyglycolic acid
PLA	Polylactic acid
PLGA	Poly-DL-lactic-co-glycolic acid
rh	Recombinant human
SEM	Standard error of the mean
SMA	α-Smooth muscle actin

1 Introduction

The loss of articular cartilage due to injury or disease can lead to pain and disability and thus have a debilitating effect on an individual. Therapeutic approaches for the regeneration of articular cartilage hold the promise of profoundly improving quality of life for many. Moreover, the lessons learned from investigating methods for the regeneration of articular cartilage will likely be of value when applied to the many other musculoskeletal connective tissues. Of importance is that there are several strategies for treating cartilage defects that have recently been introduced into the clinic, thus allowing the correlation of findings in animal models with a clinical outcome assessment. This chapter deals with specific ways in which clinical treatments may be improved though the use of modern regenerative tools associated with biomaterial scaffolds and cell therapy. Many of the principles and examples in this chapter have been drawn from recent reviews [1, 2].

The term "tissue engineering" was initially introduced to describe the technology for producing tissue in vitro [3]. More recently the term "regenerative medicine" has been used to describe the development of technology and surgical procedures for the regeneration of tissue in vivo. There are advantages and disadvantages to both strategies as they relate to the ultimate goal of regenerating articular cartilage. One advantage of the synthesis of tissue in vitro is the ready ability to examine the tissue as it forms, and to make certain nondestructive measurements to establish its functions prior to implantation. However, a disadvantage, particularly in the production of musculoskeletal tissue such as articular cartilage that must play a load-bearing role, is the absence of a physiological mechanical environment during the formation of the tissue in vitro. It is now well established that mechanical force (viz., resulting strain) serves as a critical regulator of cell function, and can profoundly influence the architecture of tissue as it is forming. Because the mechanical environment present during the formation of most musculoskeletal tissue in vivo is not well understood, it is not yet possible to recreate such an environment in vitro during the engineering of most tissues. Another disadvantage of the formation of musculoskeletal tissue outside of the body is the necessary incorporation of the tissues after implantation. This incorporation requires the engineered tissue to be mechanically coupled to the surrounding structures. Union of the implanted cartilaginous tissue with the host organ requires remodeling – degradation and new tissue formation – at the interfaces of the implant with the host tissues. That remodeling of the implanted tissue is essential for its functional incorporation.

The strategy most often taken currently for the treatment of cartilage defects is to facilitate tissue formation in vivo, under the influence of the physiological mechanical environment. However, one disadvantage of this approach is that the regenerating tissue may be dislodged or degraded by the mechanical forces normally acting at the site before it is fully formed and incorporated. This underscores the importance of post-operative rehabilitation protocols.

This chapter will deal principally with strategies directed toward the generation of articular cartilage in vivo, and therefore the term regenerative medicine will be employed throughout. Cell therapies and some tissue transplant procedures are now generally considered under the rubric of regenerative medicine because they are employed to induce regrowth of tissues that do not spontaneously grow back (i.e., that are nonregenerative). In this respect regenerative medicine is not so much a revolution in reconstructive surgery but part of the evolutionary process that this discipline has continuously undergone since its inception over 100 years ago.

It is important to point out that no tissue transplantation, tissue engineering, or regenerative medicine procedure has yet been successful in truly re-

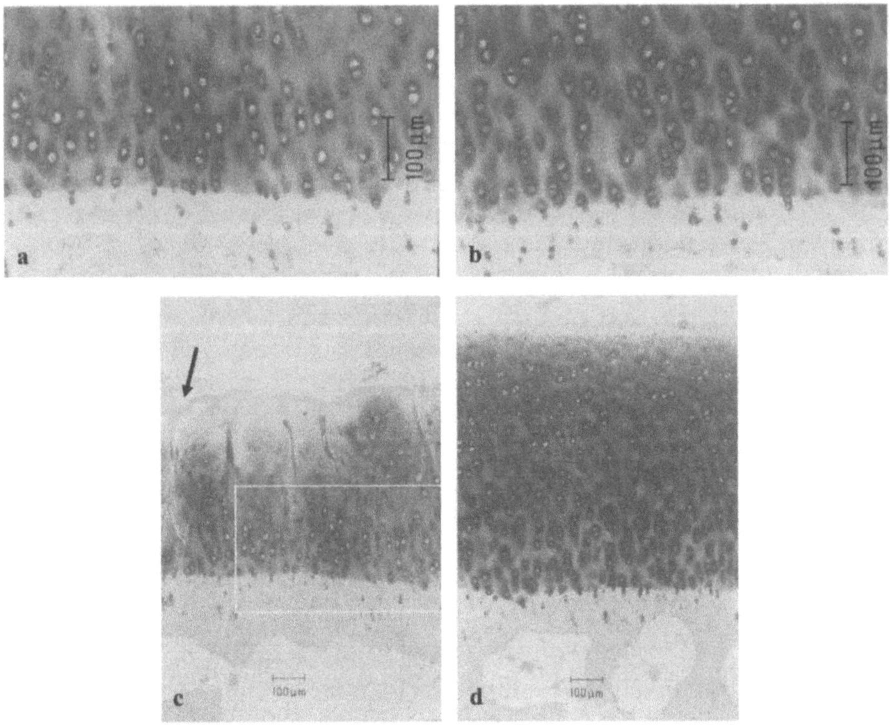

Fig. 1 **a** Histologic micrograph of an untreated chondral defect, 6 months postoperative. The region shown is that *outlined* in **c**; hematoxylin and eosin stain. **b** Histologic micrograph showing normal canine articular cartilage; hematoxylin and eosin stain. **c** Micrograph of an untreated chondral defect, 6 months postoperative. The *arrow* shows the interface between the reparative tissue on the *right* and the adjacent articular cartilage on the *left*. The *white box* shows the area presented at higher magnification in **a**; hematoxylin and eosin stain. **d** Histology of normal canine articular cartilage; hematoxylin and eosin stain

generating articular cartilage. This is of concern, as it has been known for many years that the long-term function of articular cartilage is intimately related to its composition and architecture, and to the associated mechanical properties. A promising observation, however, has been that under certain circumstances – sometimes those occurring in an untreated cartilage defect – regeneration of articular cartilage can take place, albeit only in a region of the lesion (Fig. 1). This indicates that articular cartilage regeneration is a possibility. The challenge, then, is to identify the elements of the regeneration process that need to be supplied to a particular defect: cells, matrix, cytokines, or some combination thereof.

It is also important to recognize that despite the absence of articular cartilage regeneration, many patients report a dramatic relief of pain as a result of certain treatments. This has raised fundamental questions about the criteria for success that should be adopted in the evaluation of new procedures. For example, a clinically meaningful outcome might be one that provides pain relief for 5 years, and this may have been achieved through the formation of a tissue that falls short of replicating the composition and structure of articular cartilage. Thus, is it more appropriate to use histological or clinical criteria for success of a new tissue engineering procedure? The problem with using the clinical endpoint is that patients may report a relief of symptoms for a few years, only to experience a precipitous decline in their condition. There is no reason to expect that there would be a gradual decline in function that would signal potential problems, and thus allow for adjustments to be made in the procedure or for it to be abandoned before large numbers of patients are operated on. Fundamental questions thus remain as to how to gage the success of a new cartilage repair procedure.

2
Structure and Function of Articular Cartilage

The defining histological criteria of articular cartilage are essential to recognize if the reparative tissue induced by therapeutic modalities is to be properly classified. Prior studies [4–6] have demonstrated that reparative tissue can comprise various types of cartilage (see below) as well as fibrous tissue (Fig. 2a). There are three types of cartilage that are distinguished, principally on the basis of extracellular matrix (ECM) molecular composition and appearance under the light microscope [7]. One characteristic feature of a tissue determining its classification as a cartilage is the appearance of the resident cells in a rounded morphology and contained within a lacuna [8]. The lacuna, which appears in the light microscope as a clear zone relative to the interstitial matrix, contains a finer collagen fibrillar structure as evidenced by transmission electron microscopy. Cartilages also generally contain a higher content of proteoglycans and water than other tissues.

1. Hyaline cartilage: the uniform ECM of hyaline cartilage gives a ground glass appearance in the light microscope thus explaining the use of the term "hyaline" (Fig. 2b, d). The ECM is principally comprised of type II collagen. Hyaline cartilage also has the highest water content of the cartilages, owing to its relatively high content of proteoglycans (viz., aggrecan). The following tissues are made up of hyaline cartilage: the nasal septum, tracheal rings, costal cartilages, and the epiphyseal cartilage of growing bone. The articular cartilage that lines the ends of the bones forming freely moving (i.e., diarthrodial) joints is also hyaline cartilage, but is distinguished from the other members of this class by certain architectural features described below.
2. Elastic cartilage: elastic cartilage is distinguished by its high content of elastic fibers (elastin) in addition to its content of type I collagen. It is the cartilage type found in the ear and epiglottis.

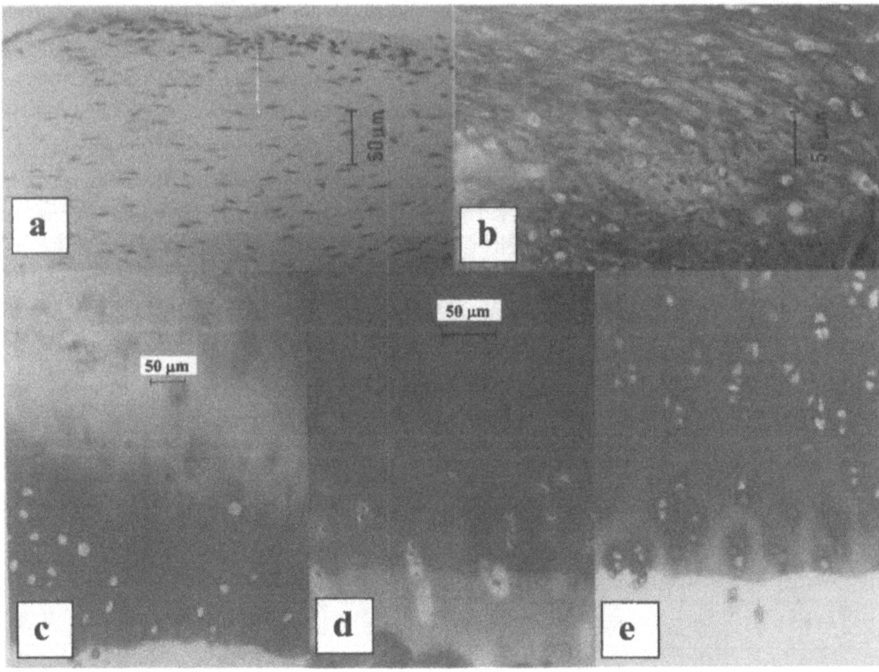

Fig. 2a–e Light micrographs of tissue types found in articular cartilage and healing defects. The micrographs have been oriented with the articulating surface at the *top* and the tidemark at the *bottom*. **a** Fibrous tissue; hematoxylin and eosin stain. The collagen fibers and elongated cells are aligned parallel to the tissue and joint surface. Untreated control after 1.5 months. **b** Fibrocartilage stained with Masson's Trichrome. Note the fibrous nature of the matrix and chondrocytic appearance of the cells. Periosteum alone group after 1 year. **c** A mixture of hyaline cartilage (*bottom* of panel) and fibrocartilage (*top* of panel). This fibrocartilage is nearly hyaline in appearance. Safranin O/fast green stain. ACI group at 6 months. **d** Hyaline cartilage stained with Safranin O/fast green. Note the complete absence of red Safranin O staining for glycosaminoglycans. ACI group at 1 year. **e** Normal articular cartilage stained with Safranin O/fast green. *Scale bars* are provided in the panels

3. Fibrocartilage: fibrocartilage is differentiated from hyaline cartilage by its fibrous appearance in the light microscope (Fig. 2b). Type I collagen is the primary collagen type in fibrocartilage. The collagen fibers often display a crimped structure like that seen in the type I collagen fibers in the ECM of ligament, tendon, and fibrous scar. Type II collagen can also be found in selected locations in fibrocartilage. Its resident cells may appear less rounded than those in hyaline cartilage, but they too appear in lacunae. Fibrocartilage comprises the menisci of joints, intervertebral discs, the pubic symphysis, and often occurs where tendon and ligament are joined to bones.

Recognition of the distinguishing features of the cartilages is essential if a reparative tissue is to be properly identified as articular cartilage. Moreover, the functionality of other types of cartilage when formed in a lesion in articular cartilage will also be based on an understanding of how the various cartilages differ.

Articular cartilage forms a layer, normally 3–4 mm thick, on the bones forming diarthrodial joints [9]. It appears whitish to the naked eye owing to its lack of vascularity and it has the feel of a dense rubber. Its function is to facilitate articulation by allowing the bones that form the joint to move freely over one another with low friction even under high joint loads. It also functions to distribute these loads uniformly to the underlying bone.

Histologically, articular cartilage is composed of chondrocytes (10% or less by volume) embedded in the ECM [10, 11]. Of importance is the fact that articular cartilage is a matrix-continuous structure, in that each cell is fully surrounded by densely packed ECM molecules. The cells cannot freely migrate through the ECM. This limitation is of particular importance as it affects the ability of the chondrocytes to migrate into sites of injury or into scaffolds implanted into defects in the tissue.

Tissue fluid contributes 60–80% of wet weight of articular cartilage [10, 11]. The high water content distinguishes articular cartilage from many other connective tissues; water makes essential contributions to the material properties and participates in joint lubrication. The structural macromolecules of the ECM account for 20–40% of wet weight. They include collagenous proteoglycans and noncollagenous proteins or glycoproteins. While type II is the principal form of collagen there are minor (but critical) amounts of collagen types V, VI, IX, and XI.

Articular cartilage is a specialized subset of hyaline cartilage that displays a distinctively organized structure [10–12] including: an arcuate collagen fiber structure originating with thick fibers rising perpendicularly from the calcified cartilage through a middle zone and bending near the superficial zone to form fibers oriented parallel with the surface; and an associated columnar arrangement of spherical cells perpendicular to the base with more flattened cells aligned parallel to the surface (Fig. 1 and Fig. 2e). In three dimensions the collagen fibrils have been described as being assembled into arcuate sheets [13]. They serve to contain the proteoglycan fraction of the ECM that in turn contains the water, which contributes to resistance to applied loads under confined compression. These structural features of the ECM, along with cell density and the pattern of proteoglycan staining, distinguish articular cartilage from other forms of hyaline cartilage (nasal and tracheal cartilage). Attachment of articular cartilage to a calcified cartilage base can be considered another distinguishing feature of this tissue type. The collagen fibrils are tethered into the calcified cartilage at the base of the tissue cartilage. The calcified cartilage is contiguous with, and mechanically coupled to, the underlying bone.

Each of the components of articular cartilage contributes to its function [12]. The chondrocytes replenish degraded proteoglycans and collagen. Proteoglycans withstand compressive loads and swell again when load is released by virtue of their ability to contain water. Collagen contains the proteoglycans and withstands tensile stresses. The anisotropy of articular cartilage structure due to the varying ECM composition and collagen orientation is reflected in several features of the tissue. For example when the surface of articular cartilage is punctured, the holes that are produced are not round, but display a slit-like feature. The orientation of the split varies with location of the cartilage surface, reflecting collagen fiber orientation. The widely varying mechanical properties of material sampled from the deep, middle, and superficial zones of articular cartilage also reflect the anisotropy of the tissue.

Articular cartilage also displays certain characteristic mechanical properties derived from the ECM composition [12, 14]. For example, when subjected to a constant load, articular cartilage undergoes time-dependent (viscoelastic) deformation, referred to as biphasic creep [15]. The creep is caused primarily by the exudation of interstitial fluid. Initially exudation occurs rapidly, as evidenced by the early rapid rate of increased deformation, and diminishes gradually until flow cessation occurs. Under high load conditions (greater than 1 MPa), about 70% of the total fluid may be squeezed from the tissue. The rate of creep is governed by the permeability of the tissue. The water expressed by the tissue participates in the fluid film lubrication of the surface. Boundary lubrication is facilitated by the adsorption of selected molecules to the cartilage surface.

That the biomechanical function of articular cartilage is so different from the other hyaline cartilages should be an indication of how important its distinctive and essential architecture are to its performance. One reason for critically distinguishing the cartilage forms that comprise reparative tissue filling in an articular cartilage defect relates to their widely differing mechanical properties and the different performance and serviceable life that could be expected as they support joint loading.

3
Limitations to Healing of Articular Cartilage

The avascular nature of articular cartilage and its relatively low cell number density of chondrocytes with low mitotic activity disadvantage healing. Defects resulting from trauma or surgery may remain indefinitely (no healing) or fill with reparative tissue comprising fibrocartilage and fibrous tissue and perhaps some hyaline cartilage, but not with articular cartilage. This is the process of repair. Regeneration of such lesions, that is, full filling with articular cartilage, while observed in a prenatal animal model [16] has never been found in the postnatal individual. In this regard articular cartilage displays some of the same healing characteristics as other connective tissues.

"Cartilage once destroyed never heals" [17, 18]. This 200-year-old observation stands as a challenge to regenerative medicine approaches to the biologic repair of injured and diseased articular cartilage. As with other tissues, cartilage depends on fibrin clot formation and the subsequent cascade of progenitor cells for its repair. The chondroprogenitor cells are derived from adjacent cartilage, underlying marrow, or synovium. Owing to the rather unique anatomical and physiological characteristics of cartilage noted above, this reparative process will not be initiated if the defect does not penetrate the subchondral bone; for the purpose of the discussion in this chapter defects contained by articular cartilage will be referred to as chondral lesions and those that penetrate the underlying bone will be referred to as osteochondral defects.

This, then, is the basis of Hunter's observation that cartilage defects do not heal. Even if a defect in the articular surface of a joint extends into the subchondral bone, patients are more likely to develop fibrocartilage of questionable functional value instead of the articular form of hyaline cartilage [19]. Thus, while healing ensues, the result is "repair" with a scar-like tissue instead of "regeneration" of articular cartilage.

Two aspects of the connective tissue healing response that are important to explore, as they may serve as a guide for regenerative medicine approaches, are cell contraction and cell migration. While these processes have been investigated for many years in dermis, it is only recently that they have been investigated in articular cartilage. Cell contraction is important for the dual role that it can play. A positive aspect of the contraction is the facilitation of healing through wound closure. A negative aspect is the organization of the fibrous scar that forms in repair and thus interferes with a regenerative response. There are examples demonstrating that strategies to reduce tissue contracture favour regeneration. An important observation supporting this inverse relationship between cell-mediated contraction and regeneration is that prenatal tissues that are regenerative do not contract. Migration is another cell process that can profoundly influence reparative processes. In order for parenchymal cells in tissue adjacent to a defect to contribute to the reparative response they need to be able to freely migrate into the provisional scaffold offered by the fibrin clot or into an implanted matrix. In articular cartilage, the resident chondrocytes are surrounded by ECM. Prior to cell division or migration, it is necessary for the cells to enzymatically degrade the surrounding matrix. The capabilities of chondrocytes to contract and migrate through their ECM have been addressed in recent studies in the context of cartilage regeneration.

3.1
Contraction of Articular Chondrocytes

It has been known for three decades that fibroblasts in selected connective tissues can express the gene for a muscle actin, α-smooth muscle actin (SMA), and contract [20, 21]. There is evidence that these cells, referred to as myofibroblasts, are responsible for dermal wound closure, and the organization of dense fibrous scar is a process that appears to interfere with regeneration [21, 22]. Up until a few years ago there was virtually no consideration of whether similar processes occurred in other connective tissues. Recent work has demonstrated that many connective tissue cells and their mesenchymal stem cell precursor can also express SMA and can contract [23]. Questions remain, however, about the specific roles of SMA-enabled connective cell contraction in normal physiological and pathological processes.

One recently completed immunohistochemical study [24] showed that a majority of chondrocytes in certain regions of human articular cartilage express SMA. The percentage of SMA-positive chondrocytes in the superficial region of human samples was 73±3% (n=22; mean±SEM) with a range of 37%–96%. In contrast, only 11±2% of the chondrocytes in the deep region of articular cartilage stained positive for this contractile isoform (range 0–29%). This was a highly statistically significant difference (Student's t-test: $p<0.0001$). These findings in human articular cartilage were supported by the results of a study of canine articular cartilage [4], which evaluated the distribution of cells containing SMA in reparative tissue resulting from spontaneous healing of surgically-created chondral defects (to the tidemark) in adult canine articular cartilage up to 29 weeks postoperatively (Fig. 3). SMA was found in a significant percentage of the cells comprising hyaline cartilage, fibrocartilage, and fibrous tissue in the defects after 6–16 weeks. Of interest was the fact that a greater percentage of cells in the cartilaginous tissues contained the contractile actin isoform, compared to the fibroblasts comprising the fibrous tissue. There was an increase in the percentage of SMA-positive cells in the fibrocartilage with time, with almost 75% of the cells containing SMA after 16 weeks. To some extent the finding of SMA-containing cells in reparative tissue in a cartilage defect parallels the initial finding of SMA-containing fibroblasts at certain stages of healing of skin wounds.

Other studies have shown that SMA-containing chondrocytes can contract a collagen-glycosaminoglycan (GAG) analog of ECM [25] and have measured the force of their contraction using a "cell force monitor" [26]. The finding of the expression of SMA was not merely due to a dedifferentiation of chondrocytes to a fibroblast phenotype. SMA-containing chondrocytes in monolayer culture continued to synthesize type II collagen and could re-adopt a spherical morphology when placed into certain environments (e.g., a type II collagen matrix) [25]. The fact that immunohistochemical staining for SMA demonstrated that articular cartilage chondrocytes in situ contained this contractile actin also served to support the supposition that SMA expression was not dependent on transdifferentiation of chondrocytes to fibroblasts (and eventually to myofibroblasts).

While studies of dermis in vivo demonstrate that contraction interferes with regeneration [27], the chondrogenic process occurring in certain models in vitro [28] appears to show a positive role for SMA-enabled contraction in the cartilage formation process. This difference in the possible positive or negative role of contraction on tissue formation processes in vitro and in vivo underscores how much we have yet to learn.

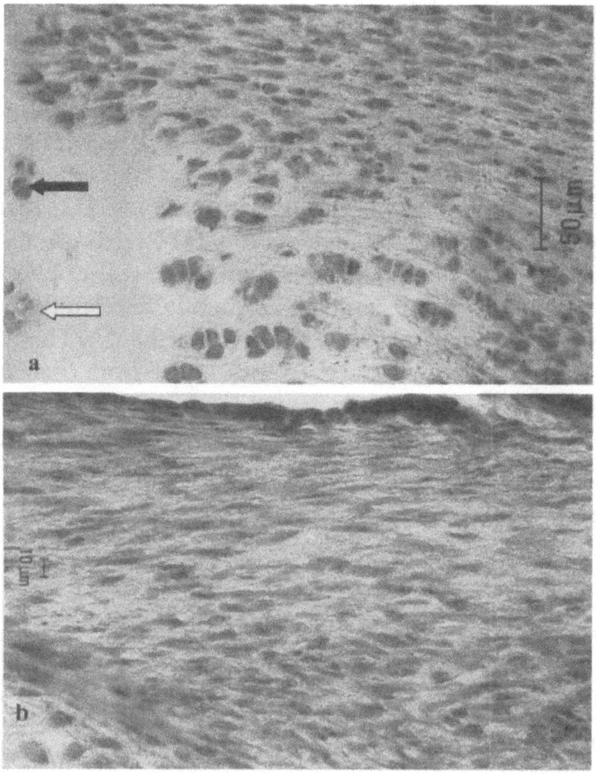

Fig. 3 a α-Smooth muscle actin immunohistochemical micrograph of a defect treated with an autologous periosteal cover alone, 6 weeks postoperatively. The micrograph demonstrates the presence of smooth muscle actin in chondrocytic cells in the articular cartilage adjacent to the defect (*black arrow*). Chondrocytes that do not contain α-smooth muscle actin are indicated by the *white arrow*. Chondrocytic and fibroblastic cells comprising the reparative tissue can be seen to contain smooth muscle actin. **b** Smooth muscle actin immunohistochemical micrograph of the central portion of a defect 6 weeks postoperative. Many of the smooth-muscle-actin-containing elongated cells in the reparative tissue were consistent in appearance with myofibroblasts

3.2
Chondrocyte Migration

That defects – fissures and larger gaps – in adult articular cartilage, which do not penetrate the underlying vascularized tissues, generally do not heal [29] reflects in some part the inability of adult articular chondrocytes to contribute to a reparative process [11]. While cloning of chondrocytes may be found in the cartilage near the site of injury [11, 30], a hypercellular zone, as seen around defects in other connective tissues, is not found bordering articular cartilage lesions. The absence of a hypercellular response may be related to the low number density and mitotic activity of the chondrocytes [31] and the absence of a fibrin clot [11]. Another factor, however, may be the inability of the cells to extricate themselves from their ECM and freely migrate through the matrix to the wound edge [11]. One prior study [32] showed the capability of isolated bovine articular chondrocytes to migrate in vitro, suggesting that the principal determinant may be the release of the cells from their ECM.

To follow up on this observation, one recent study [33] investigated the effects of enzymatic treatment on the potential for outgrowth of chondrocytes from adult human articular cartilage explants. The rate of chondrocyte outgrowth from articular cartilage could be profoundly accelerated by collagenase treatment. Moreover, many of these cells contained SMA. That under certain conditions the migration of articular chondrocytes from their matrix can be accelerated suggests that there may be conditions that could be produced in vivo to facilitate the infiltration of a provisional endogenous scaffold (viz., fibrin clot) or implanted porous matrix in a defect site with chondrocytes from the bordering cartilage.

4 Animal Models

A recent paper [34] reviewed the animal models employed for evaluating cartilage repair procedures.

4.1 Species Employed

Studies of the healing of cartilage defects have been performed in the rabbit [30, 35–40], goat [41, 42], sheep [43, 44], dog [45–48], and horse [49, 50]. The thickness of the cartilage and the chondrocyte behavior in these species differ significantly from the thickness and cellular behavior in human tissue. Moreover, all of the animal models are limited in their ability to mimic the conditions extant in the human, particular with respect to joint mechanics. Finally, the principal outcome variable in the human, pain, cannot yet be meaningfully evaluated in the animal models. These limitations notwithstanding, animal models have been of great value in the assessment of the benefits and potential problems of new cartilage repair procedures.

A principal problem of the most commonly used animal model, the rabbit, is that the defect depth is limited by the thickness of the articular cartilage in the rabbit knee (generally <0.5 mm). The thicker articular cartilage layer (approximately 0.5–1.0 mm) in the dog (Fig. 4) and goat improves this condition.

Healing may be affected by animal age, as studies done in adolescent species have reported very good healing compared with other studies [51, 52]. Although it has been known for many years that chondrocyte response to injury may be age-dependent [53], only recently has a study conclusively shown age-

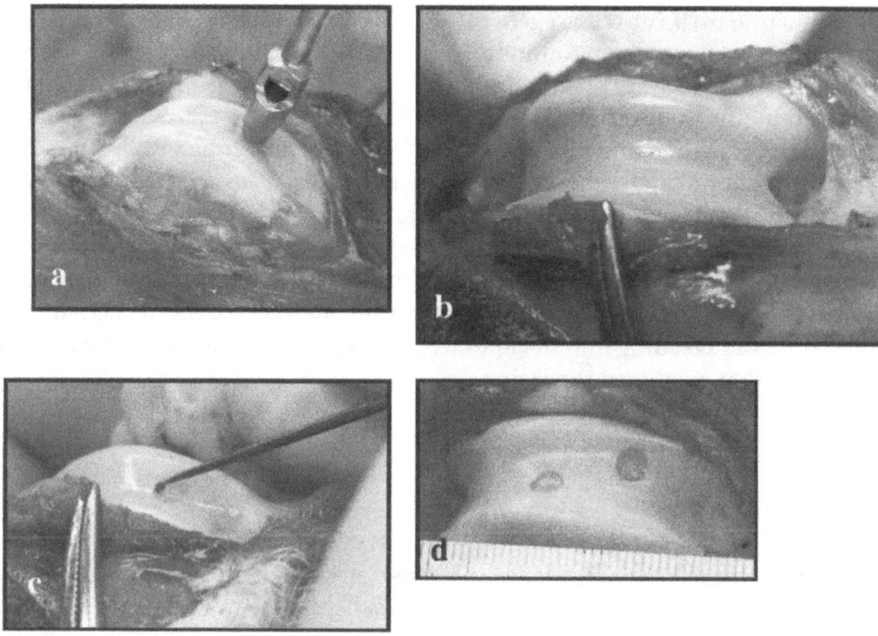

Fig. 4a–d Photographs of a chondral defect being produced in the trochlear groove of the dog. **a, b** A dermal punch is used to cut through the articular cartilage. **c** The articular cartilage is then removed using a curette under loup visualization to produce the chondral defects. **d** The thickness of the articular cartilage is approximately 0.8 mm

dependent differences in repair in the rabbit [54]. Because a treatment modality that would work only in young patients (before skeletal maturity) would be very limited in applications, adult models generally are preferred to provide more clinically applicable tests of efficacy.

4.2
The Critical Size Defect: The Effects of Depth on Healing

There is, in effect, no critical size defect – the size of defect above which no healing ensues – for articular cartilage, because even a defect as thin as a knife-edged incision may display no sign of healing (no filling with reparative tissue). The critical dimension of a defect in articular cartilage that serves as a determinant of the process of healing is depth: defects that do not penetrate the calcified cartilage versus those that do.

Defects fully contained in articular cartilage may persist with no sign of reparative tissue filling the lesion [36] or heal with limited filling by fibrous tissue, fibrocartilage, or hyaline cartilage. Remarkably, in some cases the reparative tissue that fills such (untreated) defects displays the structure of articular cartilage (Fig. 1). The positive aspect of such findings is that articular cartilage has some potential for true regeneration. The negative aspect is that such regeneration has never been found fully throughout the defect.

When healing occurs in chondral defects, the source of the cells filling such lesions remains in question. The first of the following possibilities seems the most likely: (1) blood and marrow accessing the lesion through undetected fissures through the calcified cartilage and subchondral bone; (2) cells from synovial processes growing over the intact cartilage surface to gain access to the defect; and (3) cells settling out from synovial fluid. The fibrin clot that is required to serve as the provisional scaffold to retain cells in the defect could be derived from blood from undetected fissure through the calcified cartilage (as above) or from bleeding into the joint from the synovium caused by the trauma or surgery that produced the lesion.

5 Methods of Evaluation

There are three primary methods for evaluation of the outcome of healing cartilage in an animal model [34]: histology, biochemistry, and mechanical properties. Interestingly, none of these methods match the criteria used to evaluate success clinically: pain relief, then function. Other clinical assessments that can be made through arthroscopic procedure include viewing or probing the surface. However, these results do not always correlate with patient symptoms, and are not used as reliable indicators of the success of healing.

In experimental work, most authors use histologic methods of evaluating cartilage repair. This allows evaluation of many important factors in the reparative process: the types of tissues filling the defect (including cell and ECM characteristics), attachment to adjacent structures (cartilage, calcified cartilage, or bone), and the health of the adjacent tissues. The method of staining also may be useful in eliciting biochemical information. For example several staining methods are specific to sulfated GAGs (Safranin O, alcian blue), whereas immunohistochemical stains can be used to show collagen type and cartilage specific proteins. Finally, histologic evaluation can reveal structural information, primarily collagen organization, which may give a general idea of the functionality of the reparative tissue.

A semiquantitative schema for assessing the degree of degradation of articular cartilage, the scale of Mankin et al. [53], has recently been adapted to provide quantitative assessment of the success of cartilage healing in reparative procedures [55–58]. However, this approach must be exercised with caution. The use of an ordinal semiquantitative scale precludes accurate use of parametric statistics. Still, in several studies the ordinal data are used as parametric input for the reporting of statistical comparisons [55, 59]. The meaning of these statistics should be considered only approximate. Furthermore, often a composite score of many categories is reported and used for comparison between groups. This comparison therefore is made under the assumption that the highest score possible in each category accurately reflects its relative importance in healing. This is most certainly not the case, because the "importance" in healing is at this time at best a subjective judgment of the experimenter.

More recently, quantitative histologic methods have been used for the evaluation of the reparative tissue in defects in animal models [47, 60]. In one approach, heretofore applied to rabbit investigations of osteochondral defects, the strategy has been to obtain a quantitative description of the degree of cartilage restoration: repair dimensions, degree of attachment, surface roughness and repair location [60]. The other approach [47] used to date to analyze the reparative tissue in chondral defects in a canine model has been to determine the areal percentage of selected tissue types in the defect and the percentage bonding to the adjacent articular cartilage and underlying calcified cartilage.

Other methods of analysis are more specialized and may complement histologic analysis. The measurement of mechanical properties on reparative tissue may indicate the degree to which the tissue functionally replaces normal cartilage. The major variables examined include modulus of elasticity and permeability. The biochemical analysis is usually focused on synthesis of the major components of the cartilage ECM: collagen and proteoglycans. Both of these methods have been limited in use, in part because of the destructive nature of the typical ex vivo testing procedures that prevents histologic analysis of the same tissue.

It would be desirable to have information from all of these outcomes. However, practical limitations of the research area, including the limited size of defects and expense of animal models, often make this impossible. For a preliminary investigation of healing, histologic evaluation provides the widest range of information and is widely accepted. Mechanical and biochemical evaluations are more appropriate for more specialized follow-up studies. Advances in the technology for mechanical testing, including nondestructive probes that may be used in situ, promise to expand the use of mechanical testing in analysis of cartilage repair.

6 Biomaterial Scaffolds for Regeneration of Articular Cartilage

The absence of a fibrin clot in many types of defects in articular cartilage has understandably drawn attention to the implantation of a scaffold to accommodate the infiltration of cells from surrounding tissue or as a carrier for exogenous cells. The use of three-dimensional (3-D) scaffolds that mimic the natural in vivo environment (viz., ECM) of cells has been shown to facilitate the reparative process and result in the successful growth of various functioning tissues. The 3-D environment supplied by these porous matrices serves as a desirable structural support for seeded or migrating cells and allows for a much greater surface to volume ratio for increased cell attachment as compared to a

2-D surface. There are several requirements for a scaffold to be used as an implant for articular cartilage regeneration [61]. The scaffold needs to be biodegradable, able to be fixed to the defect site, facilitate cell attachment, regulate cell expression, and possess sufficient mechanical strength [62].

Various synthetic and natural materials have been employed for the fabrication of porous absorbable scaffolds for articular cartilage tissue engineering. Among the list of absorbable or partially absorbable materials used for cartilage repair are: collagen sponge-like matrices [38, 59, 63, 64] and gels [40], hyaluronan [65], fibrin [66, 67], polylactic acid (PLA) and polyglycolic acid (PGA) [39, 68, 69], chitosan [70], devitalized cartilage [71], hydroxyapatite [72, 73], demineralized bone matrix [74, 75], and bioactive glass [62, 76]. There are advantages and disadvantages that come with using either natural or synthetic polymers as materials for these scaffolds. For example, some advantages of using synthetic polymers [such as poly-DL-lactic-*co*-glycolic acid (PLGA)] include easy molding into specific shapes, accurate control of mechanical, dissolution and degradation properties, and reproducibility. However, most synthetic polymers possess a surface chemistry that does not promote cell adhesion (which plays an important role in signal transduction pathways). Further manipulation of these synthetic polymer matrices such as coating or tethering adhesion peptides or proteins onto the surface is usually needed for sufficient cell attachment and enhanced tissue regeneration. In addition, these synthetic scaffolds (like PLGA) can produce high local concentrations of acidic by-products during degradation, which can induce an adverse inflammatory response, increase the overall degradation of the implant, or create a local environment in the scaffold that may not favor the biological activity of cells being cultured for tissue engineering purposes [77].

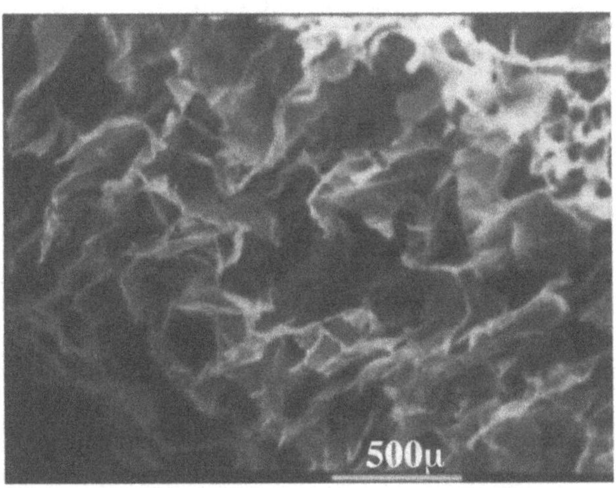

Fig. 5 Scanning electron micrograph of a type I collagen–glycosaminoglycan porous absorbable scaffold employed for tissue engineering. The material is produced by freeze-drying a blended slurry of type I collagen and chondroitin sulfate

Natural polymers such as collagen provide a more native surface to cells, since it is a major component of the natural ECM, and possesses ligands that favor cellular attachment. Collagen substrates have also been shown to modify the morphology, migration, and in some cases differentiation of cells [78]. Moreover, prior studies have demonstrated that type I collagen-GAG scaffolds produced by freeze-drying techniques (Fig. 5) can facilitate the regeneration of dermis and peripheral nerve [27, 79, 80]. Other work has demonstrated the promise of type II collagen-GAG scaffolds for articular cartilage repair [46, 48, 81]. One problem associated with natural polymer scaffolds, however, is reproducibility of structure. Numerous investigations are still ongoing to improve the properties of scaffolds for articular cartilage tissue engineering.

6.1 Use of Porous Matrices Alone (No Cells)

Several animal studies have investigated the implantation of porous, absorbable scaffolds alone in cartilage defects. However, none of these materials has yet been investigated in a clinical trial. Most studies have evaluated the reparative process in osteochondral defects implanted with these scaffold materials. Implantation of such a matrix into a chondral defect was found to have little effect in facilitating the reparative process [82]. Recent work has shown that even small penetrations through the subchondral bone, such as those produced by "microfracture", (see below) can increase the amount of reparative tissue formed in the implanted scaffold [46]. In one canine model the benefit of the matrix implanted alone in the microfracture-treated defect was to significantly increase the amount of tissue filling the defect. However, the composition of the reparative tissue was principally fibrocartilage [46]. The initial function of the scaffold appeared to be to stabilize the fibrin clot that formed as a result of the bleeding from the subchondral bone. Clinical studies will now be required to determine the extent to which implantation of a porous scaffold into a microfracture-treated site will improve the clinical outcome.

6.2 Cell-Seeded Matrices

As phenotypic changes of chondrocytes in monolayer culture have been shown in many studies, interest has focused on 3-D systems in which to culture and deliver the cells to the defect. These systems can act as templates for growth and hence contribute to phenotypic stability of the chondrocytes. The matrix, or scaffold, can play several roles in the process of tissue engineering. These roles include:

(a) Structural support for the defect site
(b) Barrier to the ingrowth of undesirable cell and tissue types
(c) Scaffold for cell migration and proliferation
(d) Carrier or reservoir of cells or regulators (e.g., growth factors)

A variety of scaffolds have shown promise thus far [61]. Ideally this scaffold should be degraded at the same speed that the cells produce their own framework. Other studies have demonstrated that matrix composition affects cell viability, cell attachment, morphology and synthesis of matrix components.

Although type I collagen matrices are the more widely researched scaffold, it was recently shown that a type II collagen matrix supported a more chondrocytic morphology and biosynthetic activity [48]. In addition to chemical composition, matrix geometry can influence the performance of the implant. Collagen can be employed as gel [64] or as a sponge with varying degrees of porosity and a range of pore diameters [48, 77, 83].

Not only is there a lack of clinical data on matrix applications for cartilage repair, there are also only a few preclinical studies in larger animals. Most of the animal work has been done in rabbits and has shown comparatively uniform good results [39, 64, 84–86].

Few studies have systematically compared different cell-seeded matrices in a large animal model. One such study conducted in dogs [82] demonstrated no notable differences in the make-up of the reparative tissue after 15 weeks in defects treated with chondrocyte-seeded type I and type II collagen scaffolds. Of importance was that a subsequent study that cultured the chondrocyte-seeded type II collagen matrix in vitro for 4 weeks prior to implantation [81] reported a greater amount of tissue filling the defect and more hyaline cartilage than in defects in which the cell-seeded construct was implanted 24 h after being seeded with cells [46]. Moreover, the construct cultured for 4 weeks also yielded more tissue, while having the same amount of hyaline cartilage, than previously found with autologous chondrocyte implantation (ACI) employed in the same animal model [5]. These results demonstrate the potential advantage of growing monolayer-expanded chondrocytes in a biomaterial scaffold in vitro prior to implantation compared to ACI.

7 Cartilage Repair Procedures Currently Employed in the Clinic

Myriad surgical procedures have recently been implemented in the clinic for the treatment of defects in the articulating surfaces of joints. These methods are directed toward accessing the subchondral vascularity and marrow either by abrasion of the calcified cartilage and subchondral bone plate with powered burrs [87–89], drilling of the subchondral bone [90], or microfracturing [91]. These procedures are usually combined with debridement of loose cartilage particles and lavage of the joint. The fibrocartilage-like reparative tissue that results from these techniques lacks the composition, structure, and mechanical properties of normal cartilage and the long-term clinical outcomes are unpredictable [92]. Patients treated with these techniques often experience short-term pain relief but develop progressive symptoms when the reparative tissue breaks down [89].

Other surgical procedures have been developed to deliver autologous chondrogenic cells to the cartilage defect in the form of a cell suspension prepared by the expansion of cells obtained from a cartilage biopsy [93] or precursor cells derived from the periosteum [94] or the perichondrium [95], with the expectation that the cells will eventually undergo terminal differentiation to chondrocytes. While these procedures have been used in selected clinics for many years, there is not yet widespread implementation. In the case of the grafting of periosteum, there have not yet been sufficient studies demonstrating reproducibility of the procedure, and in the case of perichondrium there is concern about the incidence of ossification at the implanted defect site.

Recently an alternative surgical approach has been to transplant autologous osteochondral plugs from a minimally-loaded region of the joint to the site of the lesion [96]. While the short-term results have not revealed donor site morbidity this issue demands further follow-up, particularly before the procedure is employed in patients with an otherwise healthy joint.

Of the procedures currently employed in the clinic, microfracture and the injection of autologous chondrocytes are likely the ones to benefit from tissue regeneration strategies in the near term.

7.1 Microfracture

The fact that cartilage has some limited healing potential has led to methods of improving the healing processes by increasing the number of progenitor cells able to form a cartilage-like matrix in the defect. This is accomplished by creating access to the underlying bone marrow through holes punched through the subchondral bone plate. The released marrow elements form a "super clot" which provides an enriched environment for tissue regeneration. Follow-up with long-term clinical results (based on symptomatology) of more than 8 years have been encouraging [91], despite the fact that biopsies from asymptomatic patients demonstrates the presence of principally fibrocartilaginous tissue. Because this procedure is technically the easiest one, and the least expensive, it is also the most widely utilized.

One recent study performed in adult dogs [46] investigated the implantation of a type II collagen-GAG scaffold in chondral defects (i.e., down to the tidemark -- the calcified cartilage) treated by microfracture (Fig. 6). There was significantly more reparative tissue filling microfracture-treated defects implanted with the collagen scaffold, compared to the microfracture-treated sites that did not receive implantation of the scaffolds (Fig. 7) [46]. A likely explanation for the findings was that the scaffold served to stabilize the blood clot that formed, and provided a framework for the migration of cells from the marrow cavity (possibly including mesenchymal stem cells), given access to the defect site through the microfracture holes. An important finding was that the majority of the reparative tissue in all of the microfracture-treated defects comprised fibrocartilage, as has been reported in biopsies of human cases [91].

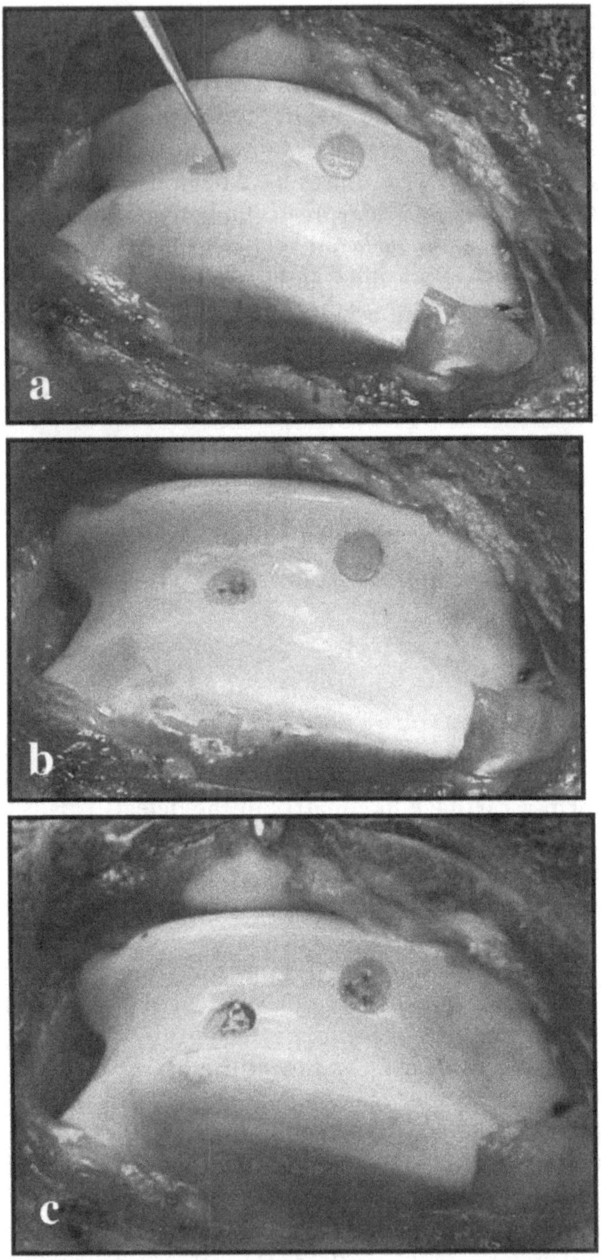

Fig. 6a–c Photographs of a pick being used to perforate the subchondral one plate in order to allow marrow and blood access to the defect site in the microfracture process

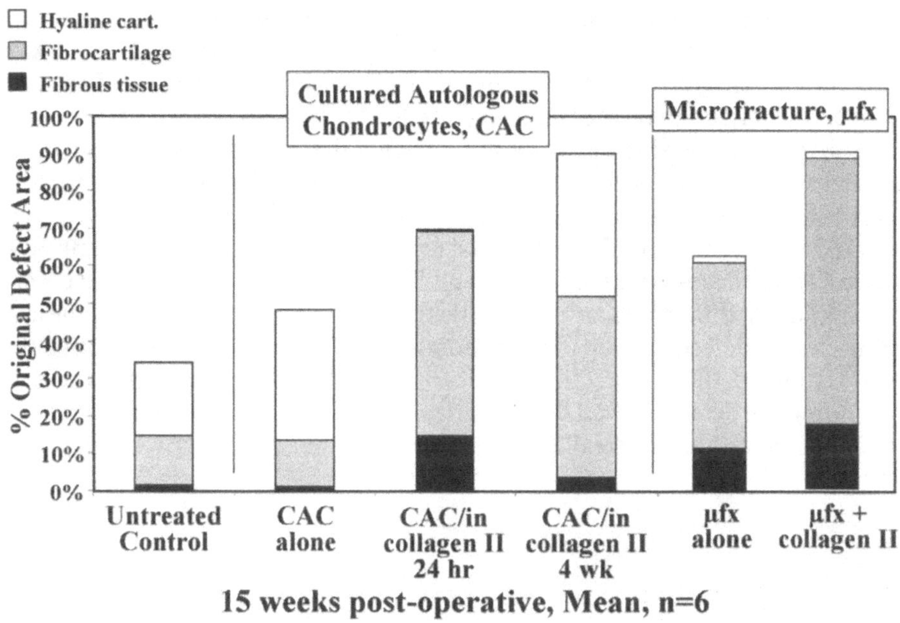

Fig. 7 Graph showing the areal percentages of the original adult canine articular cartilage defect (Fig. 4 and Fig. 6) filled with reparative tissue (the *height of the column*) and the percentages of specific tissue types comprising the reparative tissue (with articular cartilage combined with hyaline cartilage). ACI refers to autologous chondrocyte implantation. The analysis was made through a transverse section through the middle of the defect. In the *third column from the left* the chondrocytes were grown in the type II collagen-glycosaminoglycan scaffold for about 24 h prior to implantation while in the fourth column the cells were grown in the scaffold for 4 weeks prior to implantation. The difference in the amount of reparative tissue between groups was statistically significant based on analysis by ANOVA

7.2 Autologous Chondrocyte Implantation

The rationale for using autologous articular chondrocytes for a cell-based therapy is that they already possess the desired phenotype.

7.2.1 Animal Studies

Animal investigation of articular chondrocytes (viz., allogeneic cells), expanded in vitro, for cartilage repair date back more than 20 years [37]. Experiments evaluating the possibilities of ACI were first reported in 1987. In a study in rabbits Grande et al. [97] showed that chondral defects that had received transplants had a significant amount of cartilage reconstituted (82%), compared to ungrafted controls (18%). Brittberg et al. [98] later obtained similar results treating 51 New Zealand white rabbits. ACI significantly increased the amount of newly formed repair tissue up to 52 weeks in contrast to the lack of intrinsic repair with periosteal grafts alone. However, they also noted that repair tissue tended to be incompletely bonded to the adjacent cartilage.

Subsequently, Breinan et al. [5, 47] repeated these experiments in a canine model with a chondral defect (Fig. 4). They found significantly more hyaline cartilage in the ACI-treated group after 3 and 6 months compared to the untreated control. At 6 months there was a promising amount of defect filling with articular cartilage-like tissue (Fig. 7 and Fig. 8). However, by 1 year there were no significant differences among the treated and control (periosteum alone and non-treated defects) groups. By 18 months neither a complete filling, nor the restoration of the architecture was complete [47]. Moreover, cartilage surrounding the defect showed degenerative changes, some of which were related to suturing of the periosteal flap.

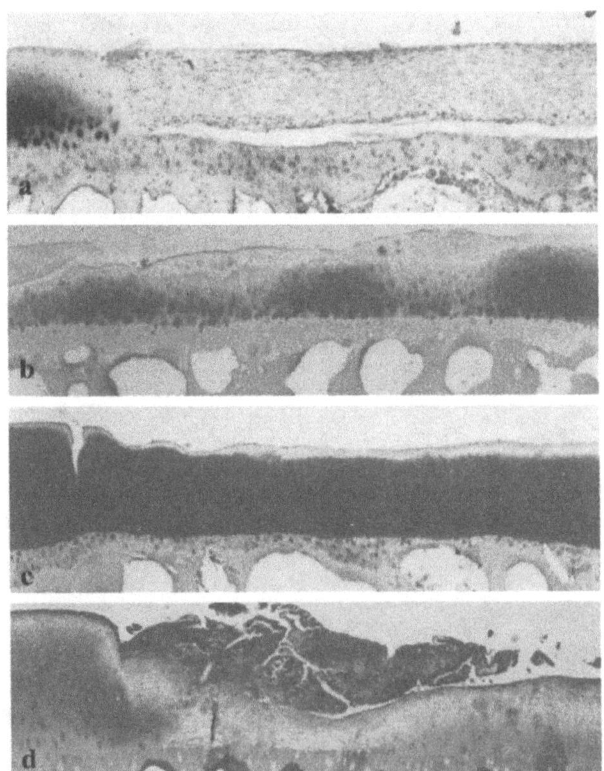

Fig. 8 Micrographs showing the histology of ACI-treated canine chondral defects (to the tidemark). Defects 1.5 (**a**), 3 (**b**), 6 (**c**), and 12 (**d**) months postoperative. (**a**) hematoxylin and eosin stain, (**b–d**) Safranin O stain

The contradictory results of ACI from various animal studies may have been due in part to differences in the animal models. Dogs have a thin subchondral bone plate that can easily be damaged. As a consequence mesenchymal stem cells can get access to the defect and mix with the implanted chondrocytes. Related work showed a significant correlation between the degree to which the calcified cartilage layer and subchondral bone were disturbed and the amount of defect filling [34], with the amount of reparative tissue inversely proportional to the remaining intact calcified cartilage. Comparing microfracture treatment with ACI showed that a greater percentage of the defect became filled with reparative tissue as a result of microfracture treatment, whereas hyaline cartilage represented a greater percentage of the reparative tissue after ACI. This work suggests the importance of an intact calcified cartilage layer for obtaining repair tissue composed mainly of articular cartilage.

Additionally, the observation that some spontaneous regeneration can occur in a canine model [4] raised the question of the degree to which such regeneration can occur in humans. In a surgically created chondral defect in adult mongrel dogs, reparative tissue filled 40% (areal percentage) of the untreated defect; 19% of the reparative tissue was articular cartilage [4]. Therefore investigations of new modalities treating lesions in articular cartilage have to acknowledge, through careful design of controls, the potential for spontaneous regeneration [4].

7.2.2 Clinical Results

On the basis of promising animal studies (rabbit) ACI was introduced into the clinic. Brittberg et al. [93] were the first to publish their results on 23 patients treated in Sweden for symptomatic cartilage defects. Thirteen patients had femoral condylar defects, ranging in size from 1.6–6.5 cm^2, due to trauma or osteochondritis dissecans. Seven patients had patellar defects. Ten patients had previously been treated with shaving and debridement of unstable cartilage. The results were very promising for the condylar defects. Patients were followed for 16–66 months (mean, 39 months). Initially, the transplants eliminated knee locking and reduced pain and swelling in all patients. After 3 months, arthroscopy showed that the transplants were level with the surrounding tissue and spongy when probed, with visible borders. A second arthroscopic examination, months after the treatment, showed that in many instances the transplants had the same macroscopic appearance as they had earlier but were firmer when probed and similar in appearance to the surrounding cartilage. Two years after transplantation, 14 of the 16 patients with femoral condylar transplants had good-to-excellent results. Two patients required a second operation because of severe central wear in the transplants, with locking and pain. A mean of 36 months after transplantation, the results were excellent or good in two of the seven patients with patellar transplants, fair in three, and poor in two; two patients required a second operation because of severe chondromalacia. Biopsies showed that 11 of the 15 femoral transplants and 1 of the 7 patellar transplants had the appearance of "hyaline-like" cartilage. These results and the fact that a commercial service for culturing autologous chondrocytes was established led to a dramatic increase in the use of this cell-based therapy for cartilage repair.

Follow-up investigations of the two largest patient groups have been reported: patients treated in Sweden predominately by L. Peterson, M.D., and M. Brittberg, M.D., and the Genzyme Cartilage Repair Registry. Peterson et al. [99] reported their 2–9 year results, including clinical, arthroscopic and histological evaluations of 101 patients. In this retrospective study ACI yielded good results in 92% for isolated femoral lesions, 67% for multiple lesions, 89% for osteochondritis dissecans and 65% for patella defects after an average follow up of 4.2 years. Arthroscopically, follow-up in 65 of 93 patients showed slow maturing of the tissue during the first year, but repair tissue at the subsequent follow-ups was as firm as the adjacent tissue. Histopathologic analysis in 21 patients revealed a homogenous matrix with low cellularity considered to be "hyaline-like" in 17 patients, whereas 4 patients showed fibrous repair tissue. Immunohistochemistry for collagen type II was positive in all of the patients with hyaline-like repair tissue, and negative in the fibrous repair tissue. Adverse events were reported in 51% of the patients, including seven graft failures (7%) and ten adhesions which needed arthroscopic intervention. Graft hypertrophy, attributed to the periosteal flap, was seen in 26 patients.

The last volume of the Genzyme Patient Registry Report included 5-year patient outcomes for individuals treated outside of Sweden. Progressive improvement in the overall condition of the patients and in symptomatology could be demonstrated at 24, 36, 48 and 60 months. Improvement compared to the baseline was 79% (78%) for all treated locations as rated by the clinician (patient). However, there was a difference between the treatment sites, with ACI being most successful at the lateral femoral condyle (100% improvement, n=3) and less successful if treating a defect in the trochlea (50%, n=2). Adverse events were reported in 7% of the patients (n=4834). These included adhesions or fibroarthrosis in 2% and hypertrophic changes in 1.3%. The cumulative incidence of treatment failure was estimated as 3.0% at 60 months. Of the patients treated, 5.9% reported re-operation following implantation. (Up-dated information is available on the web: http://www.genzymebiosurgery.com/).

7.2.3 Chondrocyte-Seeded Scaffolds

In the case of a cell therapy, a scaffold could serve as a delivery vehicle for the cells allowing for a greater number of adherent cells to be retained in the defect by virtue of the large surface area of the sponge-like matrix. Recent studies compared the reparative tissue in chondral defects in adult dogs implanted with: cultured autologous chondrocytes (CACs) alone [5], and CAC-seeded type II collagen-GAG scaffolds cultured for 24 h [46] and 4 weeks [81] prior to implantation. The cell-seeded scaffolds yielded a greater amount of reparative tissue than the sites implanted with the CACs alone (Fig. 7). The cell-seeded scaffolds cultured for 24 h induced more reparative tissue formation than the injection of cells alone, but this tissue was made up of fibrocartilage and fibrous tissue with virtually no hyaline cartilage. The question remains as to the relative importance of the amount versus the composition of the reparative tissue with respect to providing symptomatic relief for individuals with focal cartilage defects. Related to this point is the fact that the hyaline cartilage found at sites treated by CACs alone and in the collagen scaffolds did not display the architecture of articular cartilage. Of note was that the greatest amount of reparative tissue was induced by the CAC-seeded scaffold cultured for 4 weeks prior to implantation, and that this group also demonstrated the same amount of hyaline and articular cartilage (Fig. 7) as found in defects implanted with the cells alone [81]. While these studies demonstrate the promise of implementing tissue-engineering scaffolds for cartilage repair, there are potential problems and significant expense associated with the need to culture a cell-seeded scaf-

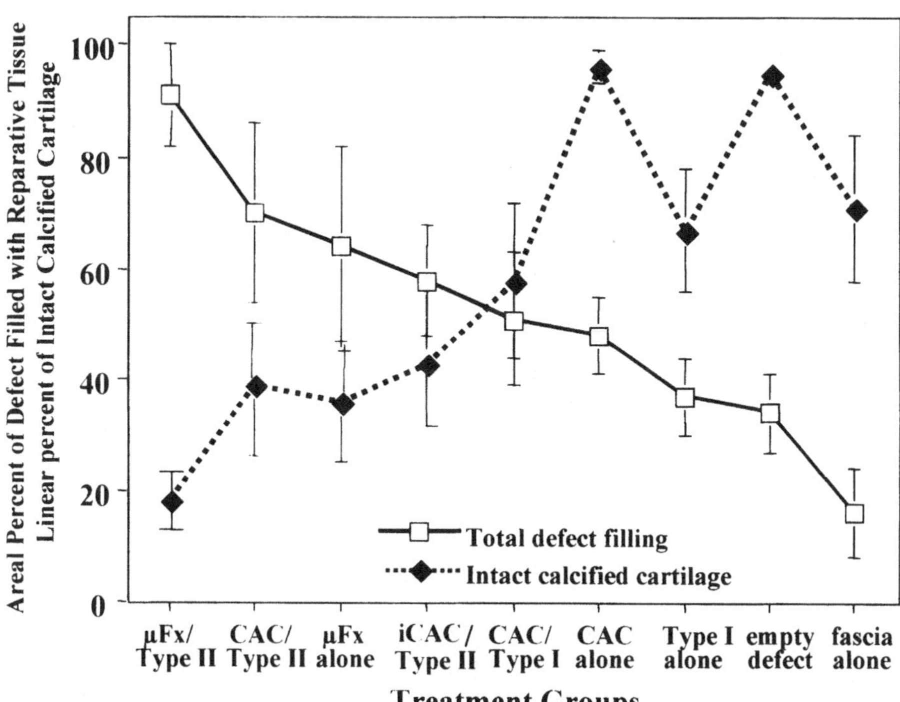

Fig. 9 Graph showing the inverse relationship between total defect filling and remaining intact calcified cartilage for 15 week canine implant groups and controls. Linear regression analysis of correlation of total filling with intact calcified cartilage (R^2=0.61). Type I and Type II refer to collagen matrices. μFx=microfracture; CAC=cultured autologous chondrocytes; iCAC=initial study of cultured autologous chondrocytes

fold for 4 weeks prior to implantation. This focuses attention on the implementation of growth factors to accelerate cell proliferation and matrix synthesis.

7.2.4 Other Cell Types for Cell Therapies

Transplantation of allogeneic osteochondral grafts has been used clinically for many years [100–102], and several investigations have focused on studying tissue engineering using allogeneic cells or cell-based tissue-engineered constructs [103, 104]. It has been proposed [64] that allogeneic chondrocytes from amputated limbs or joint arthoplasties might play a major role in the future. However, this approach is less attractive for cell therapy, because issues related to immune response and transmission of disease have to be taken into account.

The difficulty in obtaining chondrocytes and maintaining differentiated cell cultures has led to research on other cell types for cell-based therapies for cartilage repair. Several studies have shown that autologous bone-marrow-derived progenitor cells and periosteum-derived cells are able to exhibit a chondrocytic phenotype in vivo [58] and in vitro under certain conditions [105].

Friedenstein et al. [106, 107] were the first to describe the adherence of bone marrow derived stromal cells to tissue culture plastic. Using this phenotypic characteristic, he was able to easily separate mesenchymal from hematopoietic progenitor cells. Subsequently Haynesworth et al. [108], Bruder et al. [109] and Johnstone et al. [110] developed a culture system that facilitated the chondrogenic differentiation of bone-marrow-derived mesenchymal progenitor cells. Cells obtained in bone marrow aspirates were first isolated by monolayer culture and then transferred into conical tubes and allowed to form 3-D aggregates in a chemically-defined medium which included dexamethasone and/or TGF-β1. The chondrogenic differentiation of cells within the aggregate was evidenced by the appearance of toluidine blue metachromasia and the immunohistochemical detection of type II collagen [105]. Chondrogenic differentiation in this environment seemed to recapitulate embryonic chondrogenesis.

8 Role of Growth Factors and Gene Therapy

Numerous studies have shown the effects of various growth factors on chondrogenesis in vivo, and on chondrocyte proliferation, metabolism, and matrix synthesis in vitro. Among the most prominent growth factors investigated for articular cartilage tissue engineering are insulin-like growth factors (IGFs), bone morphogenetic proteins (BMPs), basic fibroblastic growth factor (bFGF, also FGF-2) and transforming growth factor-β (TGF-β). The effects of growth factors in both monolayer and 3-D culture of chondrocytes have been shown to be significant. The complexity of choosing the correct combinations and doses of growth factors to obtain the optimal tissue-engineered articular cartilage construct in vitro poses a major challenge. While there are numerous combinations of factors that can be investigated, relative to the type and dose of growth factor used, studies have already demonstrated the profound benefits of certain agents for engineered articular cartilage constructs. For example, supplementation of culture medium with IGF-1 alone has been shown to increase cell proliferation, proteoglycan synthesis, type-II collagen synthesis, and chondrogenesis, both in monolayer and in 3-D cultures [111–113].

In vivo, an improved histologic appearance and an increased proportion of type II collagen in full thickness cartilage defects in young mature horses was shown using fibrin polymers laden with IGF-1 [66]. BMPs (specifically BMP-2 and BMP-7) have also been shown to increase proteoglycan and matrix synthesis, maintain the chondrocyte phenotype, and stimulate cartilage formation in vivo in a manner similar to endochondral ossification [114]. In vivo studies using New Zealand white rabbits have demonstrated that full-thickness femoral osteochondral defects treated with rhBMP-2-supplemented collagen sponges displayed a greatly accelerated formation of new subchondral bone, an improved histologic appearance of overlying articular cartilage, and more type II collagen and tissue filling in the defect as compared with controls [115]. BMP-7 (also called osteogenic protein-1, OP-1) has also been shown to stimulate cartilage formation and aggrecan synthesis in subchondral defects in goats [116]. FGF-2 has been demonstrated to be a potent mitogen for chondrocytes and a stimulator of matrix synthesis [114]. Furthermore, in vivo studies in rabbit models [117] have reported that full-thickness cartilage defects treated with intra-articular FGF-2 had enhanced differentiation of mesenchymal cells to the chondrocyte phenotype, increased proliferation of differentiated chondrocytes, and increased accumulation of type II collagen and proteoglycan.

Growth factors can make significant contributions to cartilage repair procedures and tissue engineering by stimulating cell proliferation, migration, differentiation, and matrix synthesis. There are, however, major challenges faced in the direct application of human recombinant proteins in a clinical setting. Proteins are difficult to administer exogenously – in accurate, sustained, and therapeutically useful amounts – to sites of cartilage injury. Single bolus doses of growth factors alone in vivo have short half-lives as a result of degradation or diffusion from the defect site. Various strategies, including the use of polymers, pumps, and heparin, have been investigated as possible methods by which to achieve constant levels of growth factors at a given injured site; however, success remains limited [118]. Additionally, although it is now possible to produce large quantities of these recombinant proteins for the purpose of treatment, the expense is still another unattractive feature. A different means of supplying proteins in a localized and sustained manner in vivo is therefore needed. Delivery of a gene that could be expressed within the wound is an attractive alternative to application of the recombinant protein. Gene transfer provides the DNA that encodes for the desired protein, so that infected cells can create higher and more sustained levels of the growth factor over extended periods of time, this may be a requirement for effective articular cartilage regeneration. More than one gene can be transferred and independently regulated to supply multiple growth factors to the defect site. Some studies have even suggested that endogenously expressed proteins, induced by gene transfer, may have a more positive and more potent effect on matrix synthesis and biological activity than exogenous recombinant proteins [119].

Many questions are involved in deciding the best method of gene transfer for articular cartilage repair. Such variables include: (a) the cell, or cells, to be targeted (e.g., chondrocytes, mesenchymal stem cells, synovial cells, etc.); (b) the protein, or proteins, to be encoded; and (c) the delivery vector to be employed (which is also dependent on the size of the DNA encoding the growth factor). The vectors used in gene transfer procedures applied to articular cartilage repair include: viral vectors such as adenoviruses; lipid-mediated reagents such as liposomes; and naked DNA alone. While there have been numerous studies implementing gene transfer methods with chondrocytes in vitro, few studies have yet demonstrated a benefit for articular cartilage regeneration in vivo.

9 Summary of Factors Affecting the Healing of Cartilage Defects

Collectively, prior animal studies have demonstrated a relationship between the degree to which the calcified cartilage and subchondral plate was disrupted and the degree of defect filling and the make-up of the reparative tissue [34]. The less the disruption and remodeling of the subchondral plate, the less the filling of the defect but the greater the relative percentage of hyaline cartilage comprising the reparative tissue. These findings point to the importance of the presence of an intact calcified cartilage layer to the regeneration of articular cartilage. At the same time, however, the defects that had the greatest percentage of hyaline cartilage had the least amount of filling. The importance of the degree to which the lesion has been filled with reparative tissue with a serviceable surface layer is evidenced by the prior report that the defects filled with the greatest amount of hyaline cartilage but only 50% filling, failed to survive to 1 year post-implantation [47, 120]. It is not yet possible to determine the relative effects of the degree to which a defect is filled with reparative tissue and the composition of that tissue (fibrocartilage versus hyaline cartilage) on the function of reparative tissue in cartilage defects. Longer-term clinical follow-up of microfracture [91] and cultured ACI [93] will help to answer this question.

That the degree to which the calcified cartilage and subchondral bone is disrupted has a significant effect on the amount of reparative tissue in the defect, indicates that the blood and marrow are important sources of reparative cells and the provisional scaffold (the fibrin clot). These results also indicated an association between the degree to which the calcified cartilage layer was damaged and the type of tissue at the base of the defect. Hyaline cartilage was always found superficial to intact calcified cartilage, whereas damaged calcified cartilage was covered only by fibrous tissue or fibrocartilage. It was not determined whether the hyaline cartilage formed preferentially on intact calcified surfaces, or the formation of the hyaline cartilage played a role in healing damaged calcified cartilage. The fact that hyaline cartilage was more frequently found overlying an intact calcified cartilage layer suggests the importance of an avascular environment for the formation of this tissue type. Moreover, a calcified cartilage layer exposed by surgery may be important in facilitating the bonding of reparative tissue to the base of the defect. Formation of calcified cartilage may be necessary for incorporation of newly synthesized collagen fibers.

The finding of SMA in chondrocytes and in cells in the reparative tissue warrants further investigation. Prior studies have considered how these cells generate forces that facilitate their manipulation of the ECM to effect a tissue-specific architecture [121]. The temporal expression of α-smooth muscle actin may be responsible for these forces. Regulation of this isoform may be important to consider in cartilage repair strategies. These and other hypotheses related to the roles of SMA-enabled contraction need to be tested in future studies.

In the majority of animal models, the primary outcome variable is histology, whereas in clinical work it is pain relief. In addition, the activity level and conditions of loading can differ from the animal model to the human subject. To what extent the canine model provides an accelerated reflection of healing in the human has not yet been established. As with any animal model, the results from this work should be interpreted with caution and efforts should be made to verify them in human studies.

10 Conclusions

In recent years biologic (including regenerative medicine) therapies of cartilage defects have progressed significantly and are becoming important modalities of treatment in orthopaedic surgery. However, for all these therapies long-term outcome is unknown, and there is a lack of controlled studies comparing the different treatment options. Prospective studies are needed to better understand which of the different options will be the most suitable for specific indications.

We have to answer several questions in the near future: do we need to deliver cells to the defect or might it be suitable to recruit enough endogenous cells, locally, capable of migration, proliferation, differentiation and biosynthesis (perhaps with a method similar to microfracture)? Using in vitro tissue engineering methods, how much pre-formed matrix (in vitro tissue engineering) do we need in our construct to obtain a mechanically stable implant that will eventually integrate into the surrounding healthy articular cartilage? Is it better to have a defect fully filled with fibrocartilage or partially filled with hyaline cartilage; what will be the long-term result of these two repair tissues? And finally: is the regeneration of true (normal) articular cartilage necessary to achieve a satisfactory clinical result?

Tissue engineering holds the promise of providing effective, long-term solutions to the cartilage problems affecting a large segment of the population. While more work will be required to achieve this goal, during the course of future studies it is likely that much will be learned about cartilage biology that will inform still newer approaches for the treatment of joint problems.

Acknowledgments This work was supported in part by the Department of Veterans Affairs and the American Society for Engineering Education.

References

1. Kinner B, Spector M (in press) Cartilage – current applications. In: Goldberg VM, Caplan AI (eds) Orthopaedic tissue engineering: basic science and practice. Dekker, New York, NY
2. Capito RM, Spector M (2003) IEEE Eng Med Biol 22:42
3. Langer R, Vacanti JP (1993) Science 260:920
4. Wang Q, Breinan HA, Hsu HP, Spector M (2000) Wound Repair Regen 8:145
5. Breinan HA, Minas T, Hsu H-P, Nehrer S, Shortkroff S, Spector M (2001) J Orthop Res 19:482
6. Nehrer S, Spector M, Minas T (1999) Clin Orthop 365:149
7. Fawcett DW (1986) A textbook of histology. Saunders, Philadelphia
8. Ghadially FN (1983) Fine structure of synovial joints. Butterworths, London
9. Freeman MAR (1979) Adult articular cartilage. Pitman, Bath, Great Britain
10. Hunziker EB (1992) Articular cartilage structure in humans and experimental animals. In: Kuettner KE, Schleyerbach R, Jacques GP, Hascall VC (eds) Articular cartilage and osteoarthritis. Raven, New York, p 183
11. Buckwalter J, Rosenberg L, Hunziker E (1990) Articular cartilage: composition, structure, response to injury, and methods of facilitating repair. In: Ewing JW (ed) Articular cartilage and knee joint function: basic science and arthroscopy. Raven, New York, p 19
12. Mow VC, Ratcliffe A, Poole AR (1992) Biomaterials 13:67
13. Jeffery AK, Blunn GW, Archer CW, Bentley G (1991) J Bone Joint Surg 73-B:795
14. Armstrong CG, Mow VC (1982) J Bone Joint Surg 64A:88
15. Mow VC, Kuei SC, Lai WM, Armstrong CG (1980) J Biomech Eng 102:73
16. Namba RS, Meuli M, Sullivan KM, Le AX, Adzick NS (1998) J Bone Joint Surg 80A:4
17. Hunter W (1743) Philos Trans 470:514
18. Paget J (1853) Lect Surg Pathol (London) 1:262
19. Wirth CJ, Rudert M (1996) Arthroscopy 12:300
20. Gabbiani G, Ryan GB, Majno G (1971) Experientia 27:549
21. Schurch W, Seemayer TA, Gabbiani G (1997) Myofibroblast. In: Sternberg SS (ed) Histology for Pathologists. Lippincott-Raven, Philadelphia, PA, p 129
22. McGrath MH, Hundahl S (1982) Plast Reconstr Surg 69:975
23. Spector M (2001) Wound Repair Regen 9:11

24. Kim AC, Spector M (2000) J Orthop Res 18:749
25. Kinner B, Spector M (2001) J Orthop Res 19:233
26. Zaleskas JM, Kinner B, Freyman TM, Yannas IV, Gibson LJ, Spector M (2004) Biomaterials 25:1299
27. Yannas IV, Lee E, Orgill DP, Skrabut EM, Murphy GF (1989) Proc Natl Acad Sci USA 86:933
28. Kinner B, Zaleskas JM, Spector M (2002) Exp Cell Res 278:72
29. Buckwalter JA, Mankin HJ (1997) J Bone Joint Surg 79-A:612
30. Hunziker EB, Rosenberg LC (1996) J Bone Joint Surg 78-A:721
31. Buckwalter JA, Mankin HJ (1997) J Bone Joint Surg 79-A:600
32. Frenkel SR, Clancy RM, Ricci JL, DiCesare PE, Rediske JJ, Abramson SB (1996) Arthritis Rheum 39:1905
33. Qiu W, Meaney Murray M, Shortkroff S, Lee CR, Martin SD, Spector M (2000) Wound Repair Regen. 18:383
34. Breinan HA, Hsu H-P, Spector M (2001) Clin Orthop 391S:219
35. Amiel D, Coutts R, Abel M, Stewart W, Harwood F, Akeson W (1985) J Bone Joint Surg 67A:911
36. Fuller JA, Ghadially FN (1972) Clin Orthop 86:193
37. Green WT (1977) Clin Orthop 124:237
38. Speer DP, Chvapil M, Volz RG, Holmes MD (1979) Clin Orthop 144:326
39. von Schroeder HP, Kwan M, Amiel D, Coutts RD (1991) J Biomed Mater Res 25:329
40. Wakitani S, Kimura T, Hirooka A, Ochi T, Yoneda M, Yasui N, Owaki H, Ono K (1989) J Bone Joint Surg 71-B:74
41. Karagianes MT, Wheeler KR, Nilles JL (1975) Arch Pathol 100:398
42. Shahgaldi BF, Amis AA, Heatley FW, McDowell J, Bentley G (1991) J Bone Joint Surg 73-B:57
43. Bruns J, Kersten P, Lierse W, Weiss A, Silbermann M (1994) J Virchows Archiv 424:169
44. Rothwell AG (1990) Orthopedics 13:433
45. Altman RD, Kates J, Chun LE, Dean DD, Eyre D (1992) Ann Rheum Dis 51:1056
46. Breinan HA, Martin SD, Hsu H-P, Spector M (2000) J Orthop Res 18:781
47. Breinan HA, Minas T, Hsu H-P, Nehrer S, Sledge CB, Spector M (1997) J Bone Joint Surg 79–A:1439
48. Nehrer S, Breinan HA, Ramappa A, Young G, Shortkroff S, Louie L, Sledge CB, Yannas IV, Spector M (1997) Biomaterials 18:769
49. Hendrickson DA, Nixon AJ, Grande DA, Todhunter RJ, Minor RM, Erb H, Lust G (1994) J Orthop Res 12:485
50. Nixon AJ, Sams AE, Lust G, Grande D, Mohammed HO (1993) Am J Vet Res 54:349
51. Calandruccio RA, Gilmer WS (1962) J Bone Joint Surg 44-A:431
52. Kim HKW, Moran ME, Salter RB (1991) J Bone Joint Surg 73-A:1301
53. Mankin HJ, Dorfman H, Lippiello L, Zarins A (1971) J Bone Joint Surg 53-A:523
54. Wei X, Gao J, Messner K (1997) J Biomed Mater Res 34:63
55. Ben-Yishay A, Grande DA, Schwartz RE, Menche D, Pitman MD (1995) Tissue Eng 1:119
56. O'Driscoll SW, Salter RB (1986) Clin Orthop 208:131
57. Pineda S, Pollack A, Stevenson S, Goldberg V, Caplan A (1992) Acta Anat 143:335
58. Wakitani S, Goto T, Pineda SJ, Young RG, Mansour JM, Caplan AI, Goldberg VM (1994) J Bone Joint Surg 76-A:579
59. Frenkel SR, Toolan B, Menche D, Pitman MI, Pachence JM (1997) J Bone Joint Surg 79-B:831
60. Hacker SA, Healey RM, Yoshioka M, Coutts RD (1997) Osteoarthr Cart 5:343
61. Lee CR, Spector M (1998) Curr Opin Orthop 9:88

62. Coutts RD, Healey RM, Ostrander R, Sah RL, Goomer R, Amiel D (2001) Clin Orthop 391S:S271
63. Grande DA, Pitman MI, Alexander H (1989) Society for Biomaterials 15th annual meeting, vol 15. Lake Buena Vista, Florida, p 226
64. Kawamura S, Wakitani S, Kimura T, Maeda A, Caplan AI, Shino K, Ochi T (1998) Acta Orthop Scand 69:56
65. Solchaga LA, Dennis JE, Goldberg VM, Caplan AI (1999) J Orthop Res 17:205
66. Nixon AJ, Fortier LA, Williams J, Mohammed H (1999) J Orthop Res 17:475
67. Paletta GA, Arnoczky SP, Warren RF (1992) Am J Sports Med 20:725
68. Chu CR, Coutts RD, Yoshioka M, Harwood FL, Monosov AZ, Amiel D (1995) J Biomed Mater Res 29:1147
69. Freed LE, Marquis JC, Nohria A, Emmanual J, Mikos AG, Langer R (1993) J Biomed Mater Res 27:11
70. Nettles DL, Elder SH, Gilbert JA (2002) Tissue Eng 8:1009
71. Spangenberg KM, Peretti GM, Trahan CA, Randolph MA, Bonassar LJ (2002) Tissue Eng 8:839
72. van Susante JL, Buma P, Homminga GN, van den Berg WB, Veth RP (1998) Biomat 19:2367
73. Chiroff RT, White RA, White EW, Weber JN, Roy D (1977) J Biomed Mater Res 11:165
74. Howes R, Bowness JM, Grotendorst GR, Martin GR, Reddi AH (1988) Calcif Tissue Int 42:34
75. Reddi AH (2000) Tissue Eng 6:351
76. Suominen E, Aho AJ, Vedel E, Kangasniemi I, Uusipaikka E, Yli-Urpo A (1996) J Biomed Mater Res 32:543
77. Grande DA, Halberstadt C, Naughton G, Schwartz R, Manji R (1997) J Biomed Mater Res 34:211
78. Kleinman HK, Klebe RJ, Martin GR (1981) J Cell Biol 88:473
79. Yannas IV (1998) Wound Repair Regen 6:518
80. Chamberlain LJ, Yannas IV, Hsu HP, Strichartz GR, Spector M (2000) J Neurosci Res 60:666
81. Lee CR, Grodzinsky AJ, Hsu H-P, Spector M (2003) J Orthop Res 21:272
82. Nehrer S, Breinan HA, Ramappa A, Hsu H-P, Minas T, Shortkroff S, Sledge CB, Yannas IV, Spector M (1998) Biomaterials 19:2313
83. Frenkel SR, Toolan B, Menche D, Pitman MI, Pachence JM (1998) J Bone Joint Surg 79-B:831
84. Grande DA, Pitman MI, Peterson L, Menche D, Klein M (1989) J Orthop Res 7:208
85. Ponticiello MS, Schinagl RM, Kadiyala S, Barry FP (2000) J Biomed Mater Res 52:246
86. Dounchis JS, Bae WC, Chen AC, Sah RL, Coutts RD, Amiel D (2000) Clin Orthop 377:248
87. Friedman MJ, Berasi CC, Fox JM, DelPizzo W, Snyder SJ, Ferkel RD (1984) Clin Orthop Rel Res 182:200
88. Ewing JW (1990) Arthroscopic treatment of degenerative meniscal lesion and early degenerative arthritis of the knee. In: Ewing JW (ed) Articular cartilage and knee joint function. Basic Science and Arthroscopy. Raven, New York, p 137
89. Johnson LL (1986) Arthroscopy 2:54
90. Pridie K (1959) J Bone Joint Surg 41B:618
91. Rodrigo JJ, Steadman JR, Silliman JF, Fulstone HA (1994) Am J Knee Surg 7:109
92. Johnson LL (1991) Arthrosc J Arthrosc Rel Surg 7:14
93. Brittberg M, Lindahl A, Nilsson A, Ohlsson C, Isaksson O, Peterson L (1994) N Engl J Med 331:889
94. O'Driscoll SW (1999) Clin Orthop 367 [Suppl]:S186

95. Homminga GN, Bulstra S, Bouwmeester PSM, Van Der Linden AJ (1990) J Bone Joint Surg 72-B:1003
96. Matsusue Y, Yamamuro T, Hama H (1993) Arthroplasty 9:318
97. Grande DA, Singh IJ, Pugh J (1987) Anat Rec 218:142
98. Brittberg M, Nilsson A, Lindahl A, Ohlsson C, Peterson L (1996) Clin Orthop 326:270
99. Peterson L, Minas T, Brittberg M, Nilsson A, Sjogren-Jansson E, Lindahl A (2000) Clin Orthop 374:212
100. Ghazavi MT, Pritzker KP, Davis AM, Gross AE (1997) J Bone Joint Surg 79-B:1008
101. Gross AE, Silverstein EA, Falk J, Falk R, Langer F (1975) Clin Orthop 108:7
102. McDermott AG, Langer F, Pritzker KP, Gross AE (1985) Clin Orthop 197:96
103. Rahfoth B, Weisser J, Steinkopf F, Aigner T, von der Mark K, Brauer R (1998) Osteoarthr Cart 6:50
104. Wakitani S, Goto T, Young RG, Mansour JM, Goldberg VM, Caplan AI (1998) Tissue Eng 4:429
105. Johnstone B, Hering TM, Caplan AI, Goldberg VM, Yoo JU (1998) Exp Cell Res 238:265
106. Friedenstein AJ, Chailakhyan RK, Gerasimov UV (1987) Cell Tissue Kinet 20:263
107. Friedenstein AJ, Gorskaja JF, Kulagina NN (1976) Exp Hematol 4:267
108. Haynesworth SE, Goshima J, Goldberg VM, Caplan AI (1992) Bone 13:81
109. Bruder SP, Fink DJ, Caplan AI (1994) J Cell Biochem 56:283
110. Johnstone B, Yoo JU (1999) Clin Orthop 367 [Suppl]:S156
111. Fortier LA, Lust G, Mohammed HO, Nixon AJ (1999) J Orthop Res 17:467
112. Worster AA, Brower-Toland BD, Fortier LA, Bent SJ, Williams J, Nixon AJ (2001) J Orthop Res 19:738
113. Makower AM, Wroblewski J, Pawlowski A (1989) Cell Biol Int Rep 13:259
114. O'Connor WJ, Botti T, Khan SN, Lane JM (2000) Orthop Clin North Am 31:399
115. Sellers R, Peluso D, Morris E (1997) J Bone Joint Surg 79-A:1452
116. Louwerse RT, Heyligers IC, Klein-Nulend J, Sugihara S, van Kampen GP, Semeins CM, Goei SW, de Koning MH, Wuisman PI, Burger EH (2000) J Biomed Mater Res 49:506
117. Fujimoto E, Ochi M, Kato Y, Mochizuki Y, Sumen Y, Ikuta Y (1999) Arch Orthop Trauma Surg 119:139
118. Huard J, Li Y, Peng H, Fu FH (2003) J Gene Med 5:93
119. Smith P, Shuler FD, Georgescu HI, Ghivizzani SC, Johnstone B, Niyibizi C, Robbins PD, Evans CH (2000) Arthritis Rheum 43:1156
120. Breinan HA, Minas T, Hsu H-P, Shortkroff S, Nehrer S, Sledge CB, Spector M (1998) Tissue Eng 4:101
121. Lee GM, Loeser RF (1999) Exp Cell Res 248:294

Received: February 2004

Adv Biochem Engin/Biotechnol (2005) 94: 125–140
DOI 10.1007/b100002

Conjunctival Regeneration

Mark P. Hatton · Peter A. D. Rubin (✉)

Massachusetts Eye and Ear Infirmary, 243 Charles Street Boston, MA 02114, USA
eye_plastics@meei.harvard.edu

1 Introduction

The conjunctiva is the mucosal epithelium that lines the inner surface of the eyelids and the anterior surface of the eye. It is subject to injury from trauma or chemical burns, systemic diseases that target mucosal membranes, or as a result of eye surgery.

As with wounds throughout the body, scar formation is a normal component of conjunctival wound healing. While fibrosis is essential for wound closure, scarring of the conjunctiva may result in damage to the ocular surface that can affect vision and, in some cases, cause blindness.

This chapter provides an overview of conjunctival anatomy and physiology, provides an overview of diseases that affect conjunctiva, and reviews attempts to promote the return of normal conjunctiva after injury.

2 Anatomy

Conjunctiva consists of stratified non-keratinized epithelium and goblet cells resting on a basement membrane. The underlying substantia propria contains a rich supply of lymphocytes that aid in defense against pathogens. Conjunctiva is often considered in terms of its anatomical location, including bulbar (lining the external eye with the exception of the cornea), palpebral (lining the inner eyelid), and forniceal [junction of the bulbar and palpebral conjunctivae, forming a fornix (cul-de-sac) in the upper and lower lids] (Fig. 1). The surface area of the conjunctiva exceeds the surface area of the tissues in which it is in contact (eye and eyelid). This redundancy permits free excursion of the eye and eyelids without restriction. Conjunctiva rests on Tenon's capsule, a layer of fascia that allows for free movement of the conjunctiva over the underlying ocular surface. Conjunctiva has stronger, immobile attachments at the level of the corneal-scleral junction (limbus) and at the tarsus, a band of dense connective tissue in the eyelid.

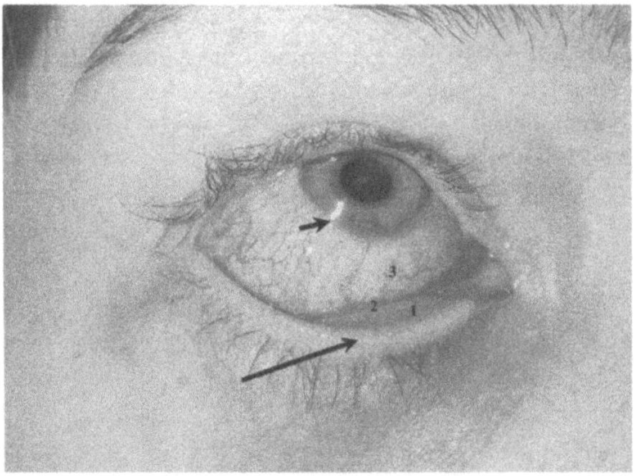

Fig. 1 The conjunctiva extends from the eyelid margin (*large arrow*) to the corneal-scleral junction, or limbus (*small arrow*). Palpebral conjunctiva lines the inner surface of the eyelid (*1*). Bulbar conjunctiva forms the outer layer of the eye (*3*). The conjunctival fornix (*2*) is located at the junction of palpebral and bulbar conjunctiva

The conjunctiva serves several important functions in the maintenance of the health of the ocular surface:

1. Physical barrier: the stratified epithelium and mucin produced by goblet cells combine to protect against infectious and mechanical trauma to the eye.
2. Lubrication: the integrity of the cornea relies on constant lubrication. The conjunctiva plays an essential role in surface lubrication by providing mucin to the tear film that aids in distributing the tear film over the eye.
3. Normal eye motility: the loosely adherent and redundant conjunctiva allows the eye to move freely and permits the eyelids to move smoothly over the eye with blinking. This aids with distributing tear film equally and preventing dryness.

The conjunctiva's close physical and functional association with the cornea has led to consideration of both tissues as forming one unit, the ocular surface. In addition to their intimate anatomical relationships, the conjunctiva and the cornea are frequently affected concurrently by injuries or systemic disease. This concept of the ocular surface is essential to understanding the goals of attempts to restore normal conjunctiva in the setting of disease and injury. As discussed below, some situations call for focal conjunctival problems to be addressed. Other disorders involve diffuse injury to the conjunctiva of one or both eyes. Finally, the co-existence of corneal and conjunctival damage requires additional interventions when options for ocular surface rehabilitation are considered.

3 Conjunctival Wound Healing

Conjunctival wound healing involves inflammation, fibroblast activation, extracellular matrix deposition and remodeling, and wound contracture [1,2]. In animal models of conjunctival wound healing, neutrophils are the predominant cell type observed one day after injury [3–5]. At day 2, the predominant cell types are macrophages and lymphocytes. Fibroblasts are recruited to the area of the defect (peaking at day 3) where they proliferate and synthesize extracellular matrix. The size of the wound is reduced by contracture owing to organization of collagen fibers combined with activity of a modified fibroblast (myofibroblast) that has contractile properties. These cells, involved in wound healing throughout the body, are located within the conjunctiva and are believed to contribute to wound closure [6]. Re-epithelialization is promoted by the constant generation of epithelial cells from stem cells followed by their migration and differentiation. These steps in healing are not unique to the conjunctiva. However, unlike other tissues throughout the body, the amount of scarring that the conjunctiva undergoes due to fibroblast activity can have severe consequences to the function of the eye and, ultimately, can cause vision loss and blindness.

Diseases that produce conjunctival inflammation have provided additional insight into the biology of conjunctival regeneration and wound healing. Inflammatory disorders that involve the conjunctiva are associated with increased numbers of T lymphocytes and macrophages within the conjunctiva [7–9]. A variety of cytokines produced by these cells are expressed at increased levels in acute conjunctival inflammation [7]. Among these, transforming growth factor-β (TGF-β) has been shown to play a critical role in the healing (and scarring) that occurs in the conjunctiva. TGF-β is a potent stimulant of proliferation of, and extracellular matrix formation by, fibroblasts throughout the body, including conjunctiva. Fibroblasts isolated from conjunctiva of patients with cicatricial pemphigoid proliferate at an increased rate compared to controls [10], which may be owing to the increased TGF-β levels in the conjunctiva of these patients.

When wound healing becomes chronic, such as in chronic inflammation, the conjunctiva undergoes a transformation in phenotype termed squamous metaplasia. This process involves loss of goblet cells and deposition of keratin on the ocular surface, which is normally populated by a non-keratinizing epithelium [11]. The result is loss of lubrication from reduction in mucin production, chronic redness, irritation, and pain. The dry eye that results from decreased mucin production, combined with mechanical irritation from keratin rubbing against the surface with each blink, leads to additional inflammation that allows the cycle to propagate. Conjunctival scarring also results in obstruction of aqueous flow from the lacrimal glands of the orbit and the accessory lacrimal glands present within the eyelid, thus contributing to surface dryness.

The epithelial tissues of the ocular surface (cornea and conjunctiva) are regenerated from stem cells. The junction of the conjunctiva and the cornea, the limbus, is the major source of stem cells that replaces the epithelium of the cornea [12–14]. Stem cells that give rise to conjunctival epithelium are located throughout the conjunctival epithelium, but are thought to be concentrated in the fornix at the junction of the palpebral and bulbar conjunctivae [15, 16], and at the junction of conjunctiva and eyelid skin [17]. Conjunctival stem cells can form both the goblet and non-goblet cell components of the conjunctival epithelium [18].

Conjunctival stem cells have the potential to migrate to the cornea and transdifferentiate to a corneal epithelial phenotype when corneal (limbal) stem cells are absent or damaged [19–24]. However, this process is imperfect, and the corneal surface may, in fact, be covered by an epithelium that has a conjunctival phenotype after injury to the limbal stem cells [25–28]. Additionally, conjunctival epithelium that develops a corneal phenotype on the cornea surface is slower to heal, demonstrates biochemical abnormalities, and has less structural integrity than native corneal epithelium [19, 20]. Replacement of corneal epithelium with conjunctiva also causes the cornea to lose one of its unique components that contribute to its optical clarity. Consequently, the replacement of corneal epithelial cells with those that have a conjunctival phenotype can impact vision severely.

Finally, when both the cornea and the conjunctiva are affected by an injury or disease, no source of stem cells exists to repopulate the ocular surface. Without intervention, such cases are associated with persistent epithelial defects, fibrosis of the ocular surface, and blindness.

4 Clinical Relevance

Approaches to restoring normal conjunctival anatomy are based on several factors:

1. Whether the disorder involves one eye or both
2. Whether the goal is restoring conjunctiva, cornea, or both
3. Whether the stem cell supply of the cornea is intact

5 Focal Conjunctival Disorders

5.1 Pterygium

A pterygium is an abnormal proliferation of conjunctival epithelial and sub-epithelial tissues that spreads to involve the cornea. It obtains its name from its characteristic wing shape (Greek "pteryx", wing). The significantly higher incidence in countries closer to the equator suggests that pterygia are caused by exposure to ultraviolet light. Histology demonstrates that the subepithelium of pterygia is replaced by fibrosis. Clinically, pterygia cause irritation, redness and distortion of the shape of the eye with the potential to decrease visual acuity. Although excision of the focal area of abnormal conjunctiva is effective in relieving symptoms, recurrence occurs in up to 60% of cases.

5.2 Post-Surgical Scarring

Many ocular surgeries require incising of the conjunctiva to gain access to deeper tissues. While small incisions often heal well, larger incisions or incisions where there is co-existing ocular inflammation may result in scarring and symblepharon formation (abnormal adhesions between conjunctival surfaces) (Fig. 2).

In some situations, the way that conjunctiva heals after surgery is critical to the outcome of the operation. For example, one option for treating glaucoma, a disease in which high pressure in the eye can cause blindness, is to create an opening in the sclera to allow fluid (aqueous) within the eye to exit. The intraocular fluid exits into the subconjunctival space where it is reabsorbed by the lymphatic system. The emanating fluid creates a visible pocket between

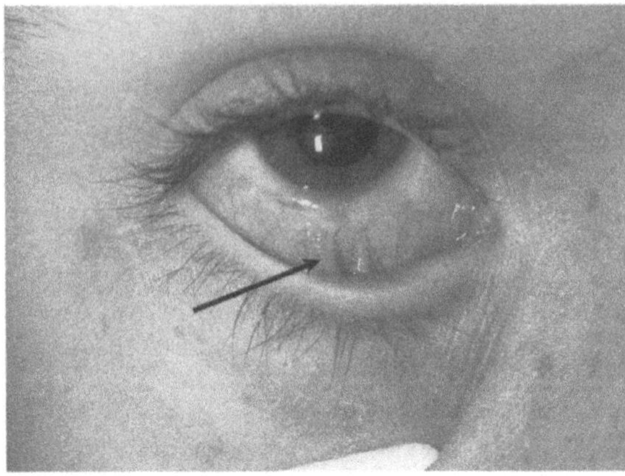

Fig. 2 Abnormal adhesions (symblepharon) between palpebral and bulbar conjunctivae following eye surgery. These scar bands can restrict the normal motility of the eye

conjunctiva and sclera where the wound is created. The normal conjunctival healing response is to create scarring and resume its position apposed to the sclera. However, when this happens after glaucoma surgery, fluid cannot leave the eye and intraocular pressure remains elevated. Attempts (discussed below) have been made to modify the course of post-surgical wound healing process in order to maintain the patency of the surgical opening.

6 Diffuse Conjunctival Damage

6.1 Conjunctival Inflammation

Inflammation of the conjunctiva is common in many diseases affecting the eye. The majority of these are self-limited (e.g., viral conjunctivitis). However, several disorders result in extensive and permanent scarring of the conjunctiva, including several immunologic disorders. Cicatricial pemphigoid is an autoimmune disease in which deposition of antibodies occurs in the basement membrane of mucous membranes, including the conjunctiva. This results in activation of complement and recruitment of lymphocytes to the affected tissues. The sequelae of the ensuing inflammation are the typical ocular manifestations seen in this potentially blinding disease: conjunctival scarring and fibrosis, symblepharon formation, altered eyelash and eyelid position, loss of goblet cells, and keratinization of the ocular surface (Fig. 3).

A similar clinical picture may be seen in Stevens-Johnson syndrome, or erythema multiforme, an immune-based disease in which desquamation of skin and mucous membranes is triggered by medications or systemic infection. Like

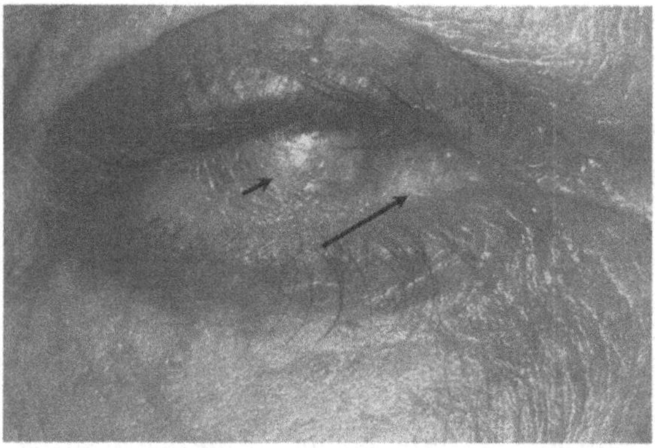

Fig. 3 Ocular cicatricial pemphigoid demonstrating severe scarring and keratinization of the cornea (*small arrow*) and symblepharon formation (*large arrow*). These patients often experience blindness from corneal opacification and severe pain due to dryness and mechanical irritation of the scarred eyelids moving across the ocular surface

cicatricial pemphigoid, the inflammation of the conjunctiva results in symblepharon formation, scarring, lid malposition, and goblet cell loss with resulting dry eye.

Trachoma, the leading cause of blindness in the world, results from infection with *Chlamydia tracomatis*, an obligate intracellular parasite. While acute infection causes only a mild conjunctivitis, chronic infection can result in severe conjunctival scarring, eyelid malposition, tear insufficiency, and corneal opacification.

6.2 Trauma

The close proximity of the conjunctiva and the cornea frequently leads to injury to both structures when a chemical splash affects the eye. While any chemical has the potential to cause severe ocular surface damage, alkali burns tend to produce the worst injuries due to their ability to penetrate through eye tissues due to saponification of cell membranes. The clinical outcome is affected primarily by the extent of injury to the stem cells. Those chemical burns that damage the entire ocular surface place the eye at greatest risk for severe corneal and conjunctival scarring.

6.3 Vitamin A Deficiency

The integrity of the ocular surface is completely dependent on constant lubrication by the tear film. Tears are composed of three layers. The aqueous layer forms the majority of the tear film and is produced by the main and accessory lacrimal glands. The lipid layer, produced by meibomian glands located within the eyelid, covers the aqueous layer and reduces evaporation of the tear film. The mucin layer is produced by conjunctival goblet cells. It aids in distributing the tear film evenly over the ocular surface. The reduction or absence of any of the three results in an insufficient or unstable tear film and the potential for breakdown of the conjunctiva and the cornea.

Vitamin A deficiency demonstrates the importance of the tear film in maintaining the integrity of the ocular surface. When vitamin A is absent, goblet cells cannot produce mucin. A severely dry ocular surface leads to keratinization of the conjunctiva and cornea. Many disorders that affect the conjunctiva (e.g., OCP, trauma) damage goblet cells. The subsequent reduction in mucin production and tear film abnormalities potentiate the ocular surface inflammation experienced by these patients.

6.4 Approaches to Promoting Conjunctival Regeneration

The majority of the published approaches to restoring normal conjunctiva are surgically based. While surgical advances certainly continue, recent insights into the biology of the ocular surface have led investigators to attempt to modify conjunctival regeneration and wound healing at the cellular and molecular levels.

6.5 Surgical Approaches

Several options exist for restoring conjunctiva surgically. The choice depends, in part on the amount of conjunctiva involved, whether the damage involves one or both eyes, and whether the goal is to restore normal conjunctiva only or promote restoration of the entire ocular surface (i.e., conjunctiva and cornea).

6.6 Conjunctival Grafting

Disease processes that are limited to a focal area of the conjunctiva can be directly excised. Excision alone and allowing the bare bed to granulate spontaneously can result in scarring and contracture. Replacing the defect with healthy conjunctiva from the same or fellow eye can prevent contracture and, in some cases, can reduce the risk of disease recurrence. For example, pterygium recurrence when it is excised may exceed 60%, while excision followed by suturing of a donor graft of conjunctiva reduces the recurrence rate to less than 5% [29]. In this technique, the donor site from the same or fellow eye can be left bare to heal spontaneously. No pterygium was present at the donor site so recurrence is not a concern at this location. Although contracture of the donor site is possible, the size of the donor graft can usually be smaller than the defect in the recipient bed owing to redundancy of conjunctiva, which allows for recipient conjunctiva to be stretched.

Conjunctival grafting has also been used to restore the defect caused by excision of ocular surface tumors and focal scarring [30, 31].

Several limitations of the autograft procedure exist. Of prime importance is that the technique is dependent on normal conjunctiva in at least one eye. Consequently it is not useful in the setting in which a disease process involves both eyes, as in bilateral chemical injury or systemic diseases such as cicatricial pemphigoid. Its main use is to restore conjunctival anatomy in the setting of focal, unilateral processes. While allograft conjunctival transplantation from living relatives has been described for disorders involving both eyes, this procedure has the risk of rejection [32, 33].

Additional limitations include the risk of damage to the graft during surgery, post-operative discomfort, and potential damage to the donor site, including fibrosis [34]. Finally, while conjunctiva has the potential to transdifferentiate to a corneal phenotype, this does not always occur, especially in the setting of chemical injury, and the process is imperfect when it does occur [16, 19, 20, 35, 36]. Re-epithelialization of the cornea based on conjunctival grafts alone may result in persistent epithelial defects, recurrent erosions, and corneal vascularization.

6.7 Replace the Conjunctiva with Non-Conjunctival Mucous Membrane

Conjunctiva that is affected in both eyes by injury or systemic disease is not suitable to serve as graft material. When reconstruction of the conjunctiva is required in the patient, alternative sources of mucosal membrane must be sought. Oral mucosa can be harvested and sewn to the ocular surface to recreate a new mucosal lining along the palpebral and bulbar surfaces after disease conjunctiva has been excised. Restoration of a mucosal-lined surface helps restore a normal anatomic relationship between the eye and eyelid and permits normal eye movement. Additionally, removal of keratinized epithelium eliminates a source of ongoing irritation to an already diseased eye. Reports of success with this method have been reported for patients with chemical injury, trachoma, and cicatricial pemphigoid [37–39].

Mucous membrane grafts do not contain stem cells. Consequently their use is limited to those situations in which stem cells have not been affected. Additionally, patients frequently report severe pain at the donor site and can develop complications including scarring that may limit movement of the lip or jaw [40]. Finally, patients may experience keratinization of the new graft and, in the case of cicatricial pemphigoid, may experience disease progression after surgery despite being controlled medically prior to grafting [39].

6.8 Amniotic Membrane

The limitations described for autologous conjunctival transfer and buccal mucosa transfer led to investigation into using non-conjunctival substrates to replace conjunctiva. Amniotic membrane, the inner layer of human placenta, is composed of a thick layer of basement membrane. Its use as a substrate that promotes wound healing is well established in other systems [41].

The first report of the use of amniotic membrane to repair conjunctival defects was published in 1940 [42]. Interest was renewed when it was demonstrated that preserved human amniotic membrane was effective in reconstructing experimentally damaged ocular surface in rabbits in which the cornea and the limbal stem cells were destroyed [27]. In these experiments, control animals (injury without amniotic membrane) developed conjunctivalization of the corneal surface. Application of amniotic membrane to the injured surface of the eye resulted in formation of an epithelium with a corneal phenotype. This suggested that the amniotic membrane promoted transdifferentiation of conjunctiva to a cornea-like epithelium, consistent with previous observations in other tissues [26, 43, 44]. These results offered promise for surface reconstruction in cases in which corneal limbal stem cells were damaged. Amniotic membrane promotes epithelialization without fibrosis or inflammation, is spontaneously absorbed, and is non-immunogenic [45, 46].

A normal conjunctival phenotype (goblet and non-goblet cell), including cell polarity, microvilli formation, epithelial stratification, and tight junction formation, develops when grown on amniotic membrane substrate [47]. Interestingly, conjunctival epithelium is repopulated only with non-goblet cells when studies are performed in vitro [47]. However, in vivo, both goblet and non-goblet cells are replaced, suggesting a role for other elements necessary for proper phenotype expression [47, 48].

Amniotic membrane is an effective substrate to replace the conjunctival defect resulting from excision of pterygia [49] and for conjunctival defects resulting from surgery to remove ocular surface tumors, symblepaharon, or scarring resulting from burns, previous eye surgery, or ocular cicatricial pemphigoid [45, 50]. In these series, the majority of patients did not develop recurrence of the original lesion and re-epithelialization occurred rapidly.

This technique has several limitations. Amniotic membrane must be harvested from human placenta. Recurrence of underlying disease, such as pterygium and scarring, may occur. When it is used for ocular surface reconstruction, it is effective at serving as a substrate to allow conjunctival tissue to re-form. However, conjunctiva does not effectively transdifferentiate to a corneal phenotype in its presence [51]. Consequently, concurrent stem cell transfer must be performed in order to restore normal corneal epithelium [52, 53]. Amniotic membrane application alone has proven unsuccessful in conditions in which severe chemical or thermal injury have affected the entire ocular surface [54] and has limited success in diffuse conjunctival inflammatory disorders such as cicatricial pemphigoid [55]. This limits amniotic membrane use for disorders in which limbal stem cells are damaged or absent unless this surgery is performed concurrently. Success for amniotic membrane transfer is greatest for replacing those lesions in which remaining conjunctiva is normal [45, 50, 53].

6.9 Other Substrates

As previously mentioned, the conjunctiva, like other epithelial cells, has the capacity to re-epithelialize spontaneously. "Anatomic" success is usually not a factor, as persistent epithelial defects of conjunctiva are rare. However, "functional" success in conjunctival wound healing is determined by the amount of fibrosis and wound contracture.

Surgical implantation of a porous collagen-glycosaminoglycan copolymer matrix at the site of experimentally induced conjunctival defects results in re-epithelialization and stroma formation without significant wound contraction [56]. In this model, epithelialization took longer in grafted than in un-grafted wounds, presumably due to the decrease in wound size in ungrafted animals that resulted from contracture. Although myofibroblasts were observed at the wound edges in grafted and non-grafted animals, they adhered to the graft in a non-ordered fashion due to the random arrangement of the collagen fibers within the matrix. In contrast, wound healing without grafting was associated with highly organized collagen and myofibroblast arrangements. It is believed that implantation of this copolymer prevented proper alignment of myofibroblasts and collagen fibers and, subsequently, wound contracture [57].

Success with an alloplastic membrane (polytetrafluoroethylene) has also been reported as a substrate after excision of conjunctival scarring [58]. This material promotes re-epithelialization beneath it rather than on its surface. After several weeks the graft is removed. Patients receiving this implant showed no scarring or wound contraction and had complete re-epithelialization.

As with amniotic membrane, the use of these implants avoids the need for harvest of donor graft from healthy tissue (conjunctiva or oral mucosa) thereby reducing operative time and morbidity. Unlike amniotic membrane they are readily available and do not require tissue harvest.

6.10 Stem Cell Transfer

The options discussed to this point assume that a supply of stem cells is present to maintain the epithelium of the cornea. In many cases, the corneal stem cells have also been damaged. While conjunctival stem cells can differentiate into a corneal epithelial phenotype and repopulate the cornea [27], this process does not always occur and the new phenotype is not perfect [16]. The corneal epithelium has its own supply of stem cells located at the limbus [12–14]. Limbal transplantation is more effective than conjunctival transplantation when the goal is reconstruction of the corneal surface [59]. This technique is effective for treating a variety of disorders of the ocular surface including those resulting from chemical burns [53], excision of tumors involving the majority of limbus, [60] and chemical injury and thermal injury [61]. However, this technique is less successful for those patients with underlying immunologic disorders such as Stevens-Johnson syndrome [53, 62, 63], who are more likely to experience infection, rejection of the graft, and corneal ulcer [62].

This technique is usually performed using autologous tissue from the fellow eye. Removing stem cells from the healthy eye places it at increased risk for poor healing [64, 65]. To avoid risking damage to the fellow eye, or when neither eye has an adequate supply of stem cells, limbal transplants may be obtained from cadaveric donors [66], but rejection of this non HLA-matched tissue may occur [63]. Grafts may be obtained from HLA-matched living donors. [33, 67] However, long-term follow-up of these patients demonstrates frequent recurrence of underlying ocular surface disease and graft rejection may still occur [67]. Systemic immunosuppression reduces the incidence of graft failure but places the patient at risk for illness associated with an immunosuppressed state [68].

7 Pharmacologic Modulation of Regeneration and Wound Healing

7.1 Anti-Metabolites

As mentioned previously, normal conjunctival wound healing can interfere with the goal of glaucoma filtering surgery. Conjunctival inflammation promotes scar formation resulting in blockage of the surgically formed drainage pathway [4]. In order to decrease conjunctival fibrosis after surgery and consequently reduce the risk of surgery failure, antimetabolites such as 5-flurouracil (5-FU) and mitomycin C have been applied in surgery or post-operatively. 5-FU inhibits thymidylate synthase, an enzyme essential for DNA synthesis and cell proliferation. Mitomycin C induces DNA cross linking and subsequent cell death.

These agents attenuate fibrosis by reducing fibroblast proliferation, migration, attachment, and extracellular matrix production as well as being able to induce apoptosis in fibroblasts [69–72]. Single, 5-min exposures to these agents are capable of inducing these effects [70, 71, 73]. This correlates with the observation in vivo that single applications of these agents result in decreased scar formation in the conjunctiva [74].

The efficacy of these agents in reducing conjunctival scarring is well documented in randomized clinical trials [75–79]. However, the use of antimetabolites is associated with a significantly increased risk of complications owing to too-low intraocular pressure and toxicity to other ocular tissues, such as the cornea [75, 76, 78, 80–82]. These complications are reduced, but not eliminated, by reducing the frequency and duration of exposure.

While exposure to anti-metabolites stops fibroblasts from proliferating, they are still capable of migration, extracellular matrix production, and growth factor expression [71, 83]. This may explain the observation that some patients treated with anti-metabolites still develop enough scarring to result in failure of the surgery [75, 76, 78, 79]. Antimetabolites have also been used to reduce the incidence of recurrence of pterygium after excision [84].

Taxol inhibits cell division by altering microtubule function and is used in the treatment of breast cancer based on its ability to inhibit cell proliferation. Taxol inhibits fibroblast proliferation and migration [85, 86]. Implantation of a slow-release carrier of taxol improved success of glaucoma filtration surgery, presumably due to inhibition of fibroblast function [87]. However, use of this agent was associated with an increased risk of serious intraocular infection [88].

7.2 Cytokines

The side effects of antimetabolites have prompted the search for other modulators of wound healing with fewer undesirable effects. For example, the identification of cytokines expressed during conjunctival inflammation and wound healing may allow for specific targeting of these molecules. TGF-β, a known fibroblast stimulant, is expressed at high levels in inflamed conjunctiva [7, 89]. This cytokine has been shown to promote fibroblast migration to wounds, extracellular matrix production, and acceleration of wound healing [90–92]. Subconjunctival injection of TGF-β induces conjunctival scarring [93] by inducing fibroblast proliferation and migration, and collagen contraction [94]. This results in increased numbers of fibroblasts and denser extracellular matrix deposition in the region of the injection [95]. Additionally, inhibitors of TGF-β result in decreased extracellular matrix production and scar formation in vivo [96].

7.3 Matrix Metalloproteinases

Matrix metalloproteinases are enzymes that cleave extracellular matrix components. These enzymes are produced by a variety of cell types, including fibroblasts, and participate in normal tissue remodeling as well as after injury. They are believed to be involved in cellular migration through extracellular matrix and participate in wound contraction throughout the body. In vitro, conjunctival fibroblasts express matrix metalloproteinases during collagen contraction [97]. Application of a broad-spectrum matrix metalloproteinase inhibitor diminishes the contraction of collagen produced by fibroblasts isolated from conjunctiva. This agent acted only on extracellular matrix remodeling and contraction and has no effect on fibroblast proliferation. Inhibition of matrix metalloproteinases also reduces conjunctival scar formation in vivo [98].

7.4 Other Approaches

Preliminary data suggest that ocular surface stem cells can be expanded in culture and returned to the ocular surface with good effect [99–101]. This offers the advantage of requiring only a small amount of tissue to be removed from the donor eye compared with the amount of tissue removed for limbal stem cell surgery. This eliminates the risk of rejection of grafts obtained from cadavers or HLA-matched relatives, and would not require immunosuppression.

Advances in gene delivery and elucidation of the molecular pathways underlying ophthalmic disease has led to ways to target ocular diseases at the level of the gene [102]. For example, the transcription factor E2F which is essential for cell proliferation can be blocked in human ocular fibroblasts by transferring "decoy" oligoneucleotides to competitively block E2F-induced transcription, with resulting inhibition of proliferation [103].

References

1. Chang L et al. (2000) Surv Ophthalmol 45:49–68
2. Khaw PT et al. (1994) Eye 8:188–95
3. Reichel MB et al. (1998) Br J Ophthalmol 82:1072–1077
4. Miller MH. et al. (1989) Ophthalmic Surg 20:350–357
5. Desjardins DC et al. (1986) Arch Ophthalmol 104:1835–1839
6. Peiffer RL, Lipper S, Merritt JC (1981) Glaucoma 3:277–280
7. Bernauer W et al. (1993) Ophthalmology 100:339–346
8. Foster CS, Rice BA, Dutt JE (1991) Ophthalmology 98:1190–1196
9. Karma A et al. (1992) Br J Ophthalmol 76:101–106.
10. Roat MI et al. (1989) Arch Ophthalmol 107:1064–1067
11. Tseng SC (1985) Ophthalmology 92:728–33
12. Schermer A, Galvin S, Sun TT (1986) J Cell Biol 103:49–62
13. Cotsarelis G et al. (1989) Cell 57:201–209
14. Tseng SC (1989) Eye 3:141–157
15. Wei ZG et al. (1995) Invest Ophthalmol Visual Sci 36:236–246
16. Wei ZG et al. (1993) Invest Ophthalmol Visual Sci 34:1814–1828
17. Wirtschafter JD et al. (1999) Invest Ophthalmol Visual Sci 40:3138–3146
18. Wei ZG et al. (1997) Invest Ophthalmol Visual Sci 38:753–761
19. Thoft RA, Friend J (1977) Invest Ophthalmol Visual Sci 16:14–20
20. Friend J, Thoft RA (1978) Invest Ophthalmol Visual Sci 17:134–139
21. Shapiro MS, Friend J, Thoft RA (1981) Invest Ophthalmol Visual Sci 21:135–142
22. Kinoshita S, Friend J, Thoft RA (1983) Invest Ophthalmol Visual Sci 24:1008–1014
23. Thoft RA, Friend J (1983) Invest Ophthalmol Visual Sci 24:1442–1443
24. Danjo S, Friend J, Thoft RA (1987) Invest Ophthalmol Visual Sci 28:1445–1449
25. Kruse FE et al. (1990) Invest Ophthalmol Visual Sci 31:1903–1913
26. Kurpakus MA, Stock EL, Jones JC (1992) Dev Biol 150:243–255
27. Kim JC, Tseng SC (1995) Cornea 14:473–484
28. Tseng SC et al. (1984) Invest Ophthalmol Visual Sci 25:1168–1176
29. Tan DT et al. (1997) Arch Ophthalmol 115:1235–1240

30. Thoft RA (1977) Arch Ophthalmol 95:1425–1427
31. Ophthalmology 89:1072–1081
32. Weise RA et al. (1985) Arch Ophthalmol 103:1736–4170
33. Kwitko S et al. (1995) Ophthalmology 102:1020–1025
34. Vrabec MP, Weisenthal RW, Elsing SH (1993) Cornea 12:181–183
35. Srinivasan BD et al. (1977) Exp Eye Res 25:343–351
36. Harris TM et al. (1985) Exp Eye Res 41:597–605
37. Ballen PH (1963) Am J Ophthalmol 55:302–312
38. Hosni FA (1974) Arch Ophthalmol 91:49–51
39. Shore JW et al. (1992) Ophthalmology 99:383–395
40. Neuhaus RW, Baylis HI, Shorr N (1982) Am J Ophthalmol 93:643–646
41. Trelford JD, Trelford-Sauder M (1979) Am J Obstet Gynecol 134:833–845
42. De Rotth A (1940) Arch Ophthalmol 23:522–525
43. Guo M, Grinnell F (1989) J Invest Dermatol 93:372–378
44. Streuli CH, Bailey N, Bissell MJ (1991) J Cell Biol 115:1383–1395
45. Tseng SC et al. (1998) Arch Ophthalmol 116:431–441
46. Schwab IR, Isseroff RR (2000) N Engl J Med 343:136–138
47. Meller D, Tseng SC (1999) Invest Ophthalmol Visual Sci 40:878–886
48. Meller D, Dabul V, Tseng SC (2002) Exp Eye Res 74:537–545
49. Prabhasawat P et al. (1997) Ophthalmology 104:974–985
50. Tseng SC, Prabhasawat P, Lee SH (1997) Am J Ophthalmol (1997) 124:765–774
51. Cho BJ et al. (1999) Cornea 18:216–224
52. Tsubota K et al. Surgical reconstruction of the ocular surface in advanced ocular cicatricial pemphigoid and Stevens-Johnson syndrome. Am J Ophthalmol (1996) 122:38–52
53. Gomes JA et al. (2003) Arch Ophthalmol 121:1369–1374
54. Joseph A, Dua HS, King AJ (2001) Br J Ophthalmol 85:1065–1069
55. Barabino S et al. (2003) Ophthalmology 110:474–480
56. Hsu WC et al. (2000) Invest Ophthalmol Visual Sci 41:2404–2411
57. Yannas IV (1998) Wound Repair Regen 6:518–523
58. Levin PS, Dutton JJ (1990) Arch Ophthalmol 108:282–285
59. Tsai RJ, TT Sun TT, Tseng SC (1990) Ophthalmology 97:446–455
60. Copeland RA Jr, Char DH (1990) Am J Ophthalmol 110:412–415
61. Kenyon KR, Tseng SC (1989) Ophthalmology 96:709–22; discussion 722–723
62. Samson CM et al. (2002) Ophthalmology 109:862–868
63. Tsubota K et al. (1999) N Engl J Med 340:1697–1703
64. Chen JJ, Tseng SC (1991) Invest Ophthalmol Visual Sci 32:2219–2233
65. Chen JJ, Tseng SC (1990) Invest Ophthalmol Visual Sci 31:1301–1314
66. Tsai RJ and Tseng SC (1994) Cornea 13:389–400
67. Rao SK et al. (1999) 106:822–828
68. Daya SM, Ilari FA (2001) Ophthalmology 108:126–33; discussion 133–134
69. Lee DA et al. (1990) Invest Ophthalmol Visual Sci 31:2136–2144
70. Occleston NL et al. (1994) Invest Ophthalmol Visual Sci 35:3681–3690
71. Occleston NL et al. (1997) Invest Ophthalmol Visual Sci 38:1998–2007
72. Yamamoto T et al. (1990) Ophthalmology 97:1204–1210
73. Khaw PT et al. (1992) Arch Ophthalmol 110:1150–1154
74. Bergstrom TJ et al. (1991) Arch Ophthalmol 109:1725–1730
75. The Fluorouracil Filtering Surgery Study Group (1989) Am J Ophthalmol 108:625–635
76. The Fluorouracil Filtering Surgery Study Group (1996). Am J Ophthalmol 121:349–366
77. Honjo M et al. (1998) Am J Ophthalmol 126:823–824
78. Katz GJ et al. (1995) Ophthalmology 102:1263–1269

79. Skuta GL et al. (1992) Ophthalmology 99:438–444
80. Suner IJ et al. (1997) Ophthalmology 104:207–214; Discussion 214–215
81. Knapp A et al. (1987) Am J Ophthalmol 103:183–187
82. Rubinfeld RS et al. (1992) Ophthalmology 99:1647–1654
83. Daniels JT et al. (1999) Exp Eye Res 69:117–127
84. Chen PP et al. (1995) Am J Ophthalmol 120:151–160
85. van Bockxmeer FM et al. (1985) Invest Ophthalmol Visual Sci 26:1140–1147
86. Joseph JP, Grierson I, Hitchings RA (1989) Curr Eye Res 8:203–215
87. Jampel HD et al. (1993) Invest Ophthalmol Visual Sci 34:3076–3083
88. Jampel HD, JI Moon (1998) J Glaucoma 7:170–177
89. Elder MJ, Dart JK, Lightman S (1997) Exp Eye Res 65:165–176
90. Grant MB et al. (1992) Invest Ophthalmol Visual Sci 33:3292–3301
91. Brown GL et al. (1988) Ann Surg 208:788–794
92. Ishikawa O et al. (1990) Biochem Biophys Res Commun 169:232–238
93. Cordeiro MF et al. (1999) Invest Ophthalmol Visual Sci 40:1975–1982
94. Cordeiro MF et al. (2000) Invest Ophthalmol Visual Sci 41:756–763
95. Doyle JW et al. (1997) Invest Ophthalmol Visual Sci 38:1630–1634
96. Cordeiro MF, Gay JA, Khaw PT (1999) Invest Ophthalmol Visual Sci 40:2225–2234
97. Daniels JT et al. (2003) Invest Ophthalmol Visual Sci 44:1104–1110
98. WongTT, Mead AL, Khaw PT (2003) Invest Ophthalmol Visual Sci 44:1097–1103
99. Pellegrini G et al. (1997) Lancet 349:990–993
100. Shatos MA et al. (2003) Invest Ophthalmol Visual Sci 44:2477–2486
101. Shatos MA et al. (2001) Invest Ophthalmol Visual Sci 42:1455–1464
102. Chaum E, Hatton MP (2002) Surv Ophthalmol 47:449–469
103. Akimoto M et al. (1998) Exp Eye Res 67:395–401

Received: February 2004

Adv Biochem Engin/Biotechnol (2005) 94: 141–179
DOI 10.1007/b100003

Heart Valve Regeneration

Elena Rabkin-Aikawa[1] · John E. Mayer Jr.[2] · Frederick J. Schoen[1] (✉)

[1] Department of Pathology, Brigham and Women's Hospital, 75 Francis Street, Boston, MA 02115, USA
fschoen@partners.org

[2] Department of Cardiovascular Surgery, Children's Hospital, Harvard Medical School, Boston, MA, USA

Abstract The valves of the heart cannot regenerate spontaneously. Therefore, heart valve disease generally necessitates surgical repair or replacement of the diseased tissue by mechanical or bioprosthetic valve substitutes in order to avoid potentially fatal cardiac or systemic consequences. Although survival and quality of life is enhanced for many patients treated surgically, currently available valve substitutes remain imperfect. This is especially the case in pediatric applications, where physiologically corrective procedures can be successfully performed, but reoperations are frequently required to replace failed valve substitutes or accommodate growth of the patient. While much work is currently underway to incrementally improve existing valve substitutes, a major impact will require radically new technologies, including tissue engineering or regeneration. The use of engineered tissue offers the potential to create a non-obstructive, non-thrombogenic tissue valve substitute containing living cells capable of providing ongoing remodeling and repair of cumulative injury to the extracellular matrix. Ideally, this would allow growth in maturing recipients. The innovative fabrication of materials and the development of sophisticated methods to repair or regenerate damaged or diseased heart valves requires integration of a diverse array of basic scientific principles and enabling technologies. Thus, heart valve tissue engineering requires an understanding of relationships of structure to function in normal and pathological valves (including mechanisms of embryological development, tissue repair and functional biomechanics), and the ability to control cell and tissue responses to injury, physical stimuli and biomaterial surfaces, through chemical, pharmacological, mechanical and potentially genetic manipulations. These approaches created by advances in cell biology raise exciting possibilities for in situ regeneration and repair of heart valves.

Keywords Heart valve · Regeneration · Valvular interstitial cells · Valvular endothelial cells · Tissue engineering

Abbreviations and Symbols

ECM	Extracellular matrix
TGF	Transforming growth factor
ErbB 2	Human epidermal growth factor receptor 2
eNOS	Endothelial isoform of nitric oxide synthase
NF	Nuclear factor
NFAT	Nuclear factor of activated T-cells
EGFr	Epidermal growth factor receptor
Shp2	Phospho-tyrosine phosphatase
MHC	Major histocompatibility complex
VEC	Valvular endothelial cells
VIC	Valvular interstitial cells
GAG	Glycosaminoglycan
vWF	von Willebrand's factor
PECAM-1	Platelet-endothelial cell adhesion molecule-1
VEGF	Vascular endothelial growth factor
VCAM-1	Vascular cell adhesion molecule-1
ICAM-1	Intracellular adhesion molecule-1
MMP	Matrix metalloproteinase
TIMP	Tissue inhibitor of matrix metalloproteinase
SM	Smooth muscle myosin heavy chain isoform
SMA	Smooth muscle actin
SMemb	Nonmuscle myosin heavy chain isoform
NAV	Normal aortic valve
NPV	Normal pulmonary valve
PGA	Poly(glycolic acid)
PLLA	Poly(L-lactic acid)
PLGA	Poly(lactic-*co*-glycolic acid)
PHA	Polyhydroxyalkanoate
TEHV	Tissue-engineered heart valve

1 Introduction

Disease of the heart valves commonly leads to valve dysfunction. Since the tissue of the heart valves cannot regenerate spontaneously, valve disease generally necessitates surgical repair or replacement of the diseased tissue in order to avoid serious and potentially fatal cardiac or systemic consequences. Surgical replacement of diseased human heart valves by mechanical and bioprosthetic valve substitutes is now commonplace, with approximately 85,000 valve replacements done each year in the US and 275,000 worldwide. Although survival and quality of life is enhanced for many patients so treated, the therapies presently available not only provide imperfect functional restitution, but also have potential complications. Indeed, currently available mechanical and tissue valve substitutes remain imperfect [1]. The essential requirements for an efficacious and safe heart valve replacement device are summarized in Table 1.

Four categories of valve-related complications predominate: (a) thromboembolism, thrombosis, and anticoagulation-related hemorrhage; (b) prosthetic valve endocarditis (i.e., infection); (c) structural dysfunction (i.e., degeneration of the prosthesis biomaterials); and (d) nonstructural dysfunction (e.g., tissue overgrowth, paravalvular leak, hemolysis and other miscellaneous problems such as entrapment of a moving prosthesis part by a suture or noise). The overall rate of problems is similar for mechanical valves and tissue valves, but they differ in their major causes of failure [2, 3]. The combined risk of thromboembolic complications and hemorrhage (the latter owing to the anticoagulation therapy required to prevent thromboembolism) constitutes the principal disadvantage of mechanical prosthetic valves. However, contemporary mechanical valves have high structural reliability. In contrast, the advantages of tissue valves include a central pattern of flow resembling that of native cardiac valves and low thrombogenicity, so that most patients with bioprosthetic valves do not need long-term anticoagulation. However, tissue valves frequently suffer from tissue degradation with tearing and/or calcification of the valve leaflets [4]. These problems are a direct result of inherent and fundamental deficiencies of

Table 1 General requirements for a substitute heart valve

Non-obstructive
Closure is prompt and complete
Non-thrombogenic
Infection-resistant
Chemically inert and non-hemolytic
Durable
Easily and permanently inserted
Not annoying (noise-free)

Modified from [130].

existing bioprostheses: their failure to repair and remodel in an environment of repetitive stress and strain on the leaflets. Moreover, and especially germane to the pediatric population, contemporary heart valve substitutes are unable to grow with the patient. Indeed, pediatric patients requiring valve replacement will usually need multiple operations to place successively larger devices to accommodate growth; thus repetitive surgical risk will be additive to the other typical post-operative device complications.

The "holy grail" is to create or regenerate a living valve replacement that functions well hemodynamically, repairs ongoing tissue damage, and has long-term durability and growth potential similar to those of the natural heart valves. Owing to the complexity of this goal, an unprecedented integration of biological and engineering science and technology will be required for successful preclinical developmental studies and clinical trials. In addition, innovative regulatory approaches and commercialization strategies will be needed. This chapter strives to frame this discussion; we will review concepts and current data on normal heart valve structure, mechanics, physiology, and development, and provide a progress report on the emerging field of heart valve tissue engineering. We will emphasize both the exciting potential of this technology and the key barriers that will have to be overcome before clinically useful engineered heart valves can be realized.

2 Pathologic Considerations in Heart Valve Disease

Dysfunction of the native cardiac valves has several overall causes [5]. Most frequently, intrinsic disease of the valve leaflets/cusps results from leaflet or cuspal calcification, fibrosis, fusion, retraction, perforation, rupture, or stretching. Secondly, there can be congenital valvular malformations, often involving associated structures. Thirdly, there may be dilation of, damage to or distortion of the supporting structures without primary leaflet or cuspal pathology. Two functional effects of valve disease are possible: inhibition of forward flow secondary to obstruction caused by failure of a valve to open completely (known as *stenosis*), and reverse flow caused by failure of a valve to close completely (known as *insufficiency*). Calcific aortic valve disease, owing to dystrophic calcification of the aortic valve cusps and ring (called the annulus), is the most common cause of aortic stenosis, the single most common clinically important valve problem. The leading cause of chronic aortic insufficiency is aortic root dilation, causing stretching and outward bowing of the commissures and impaired cuspal coaptation. Deformity becoming limiting many years subsequent to rheumatic fever is the leading cause of mitral stenosis. Degenerative mitral valve disease (mitral valve prolapse, also known as myxomatous mitral valve) and mitral regurgitation secondary to ischemic heart disease are the leading causes of pure mitral valve regurgitation. The most common causes of dysfunction of the pulmonary and tricuspid valves are congenital malformations.

In addition, the function of any of the four cardiac valves can be affected by infectious processes (called endocarditis), which are capable of destroying valve tissue in days to weeks and usually result in valvular insufficiency.

3
Cardiac Valve Development

During normal development of the heart, the valve cusps/leaflets originate from mesenchymal outgrowths known as endocardial cushions [6, 7, 131, 132]. The cells arise from several sources: neural crest cells, endocardial cells, and possibly cells of epicardial origin. Neural crest cells initially envelop the endothelium of the caudal arteries then extend into the aortic sac and initiate the formation of the aorticopulmonary septum. The cardiac neural crest also contributes to the smooth muscle cell media of the blood vessels and the semilunar (ventricular outflow) valves (aortic and pulmonary) [8]. Endocardial cushion formation involves a subset of endothelial cells in the cushion-forming area that initially line the internal cardiac surface but undergo endothelial to mesenchymal transformation (called "transdifferentiation") and migrate from the luminal surface to the subendocardial tissue. Thus, newly formed mesenchymal cells migrate into the underlying extracellular matrix (ECM) [9–11, 133], and remodel it to transform the cushions into leaflets and cusps (Fig. 1). A biochemical marker for transdifferentiation is the expression of α-smooth muscle actin, which is not normally expressed by endothelial cells [12].

Complex mechanisms regulate the epithelial-to-mesenchymal transition that occurs in development and pathologic states at multiple anatomic sites [13, 14]. Developmental studies have shown that transforming growth factor (TGF)-β2 and TGF-β3 promote transformation of endocardial cells into mesenchymal cells in the heart [15]. It has been proposed that the binding of TGF-β2 to its receptor TBR-II promotes TBR-I and TBR-II signaling complex to activate downstream mediators that result in atrioventricular cushion formation [16]. Activation of ErbB receptor is important for induction of cardiac endothelial cell transformation by the ECM hyaluronan-modulated pathway [17]. Periostin, a cell adhesion protein, may also be important in the process of cardiac valve development [134].

Studies in mice have revealed numerous other molecular components, which are either present during or are essential for normal morphogenesis of valve leaflets. For example, mice lacking functional eNOS demonstrate a high incidence of congenitally deformed (bicuspid) aortic valves [18]. Inactivation of NF1 in mice leads to midgestation lethality from cardiovascular abnormalities, including structural malformation of the outflow tract of the heart and enlarged endocardial cushions [19]. Other genetic knockouts with defective valvulogenesis and congestive heart failure include endoglin, Smad6, neurofibromin-1 and NFATc1 [20]; mice in which the transcription factor NFATc1 has been knocked out fail to form aortic and pulmonary valves [21], whereas the rest of the vasculature forms normally. Mice mutants for epidermal growth factor receptor EGFr and Shp2 have also been found to have defects in valvulogenesis [22]. Such studies suggest that unique signaling pathways exist in the endothelium of developing heart valves, though whether these pathways affect the phenotype and function of valvular endothelium in the post-natal valve remains to be investigated. Some studies suggest that the process of transdifferentiation of endothelial cells into interstitial cells may persist post-natally.

The evolution of connective tissue components and phenotypes of the cellular elements of heart valves during further maturation in utero and the subsequent changes that occur at birth, when intracardiac pressure-volume relationships change drastically, are largely unknown.

Figures 1A–C see page 147–149

Fig. 1A–C Valve development-transdifferentiation. **A** schematic diagram of valvuloseptal morphogenesis of the developing heart. The primordial of valvuloseptal tissue are the endocardial cushions. The JB3 progenitors of the cushion cell originate from the heart-forming fields and may also give rise to the myocardial lineage. Following their differentiation into pre-endocardial mesenchyme, these cells coalesce - possibly along with vascular mesenchyme that originates from outside the heart-forming fields - to form the endocardial endothelium. The cushions first develop as regionalized swellings of the cardiac jelly and contain protein complexes of fibronectin and ES proteins (adherons). These ES proteins, produced by the myocardium, elicit the transformation of endothelial cells to cushion mesenchyme. Differentiation of the cushions into valvuloseptal tissue is mediated, in part, by the transitional matrix, which includes the microfibririllar proteins fibulin and fibrillin. **B** Depiction of the molecular events involved in the transformation of endothelial cells into cushion mesenchyme. The myocardium of the primitive heart tube demonstrates a segmental pattern of molecular expression. Molecules expressed uniquely within the AV canal and ventricular outflow tract regions include transforming growth factor (*TGF*)-β2, bone morphogenic protein (*BMP*)-4, *msx-2*, and the ES proteins. These latter proteins form complexes with fibronectin and appear to regulate endothelial cells' transformation. As endothelial cells are activated to transform, cell surface neural cell adhesion molecule (*N-CAM*) is downregulated as the cells begin secreting higher levels of both substrate adhesion molecules (*SAM*), i.e, cytotactin, and proteases. Also, cells undergoing transformation increase their expression of TGF-β and *msx-1* homeobox gene. Subsequently, cells that have converted to mesenchyme express the helix–loop–helix protein Id and begin to produce the ECM proteins fibulin and fibrillin. **C** Diagrammatic summary of endocardial cytodifferentiation. In the *left panel*, the endocardium of an 8 somite embryo is shown as consisting of uniformly similar cell types in all areas of the simple tubular heart. In the middle panel, the two pathways of endocardial development are shown as they first occur in 16–18 somite embryos: in the atrium and ventricle, the endocardium becomes attenuated with reduced secretory potential and acquires an "endothelial look". Conversely, in sites where prospective valvular and septal primordial (cushion tissue) will form, the endocardium hypertrophies and acquires an amplified secretory potential. The initial formation of cushion tissue cells beneath the endocardium (*right panel*), and their morphologic similarity to the latter, suggest that valvular interstitial cells are endocardial derivatives. **A, B** reproduced by permission from [7]. **C** Reproduced from [126], with permission from Academic Press, Orlando, Florida and by permission from [10]

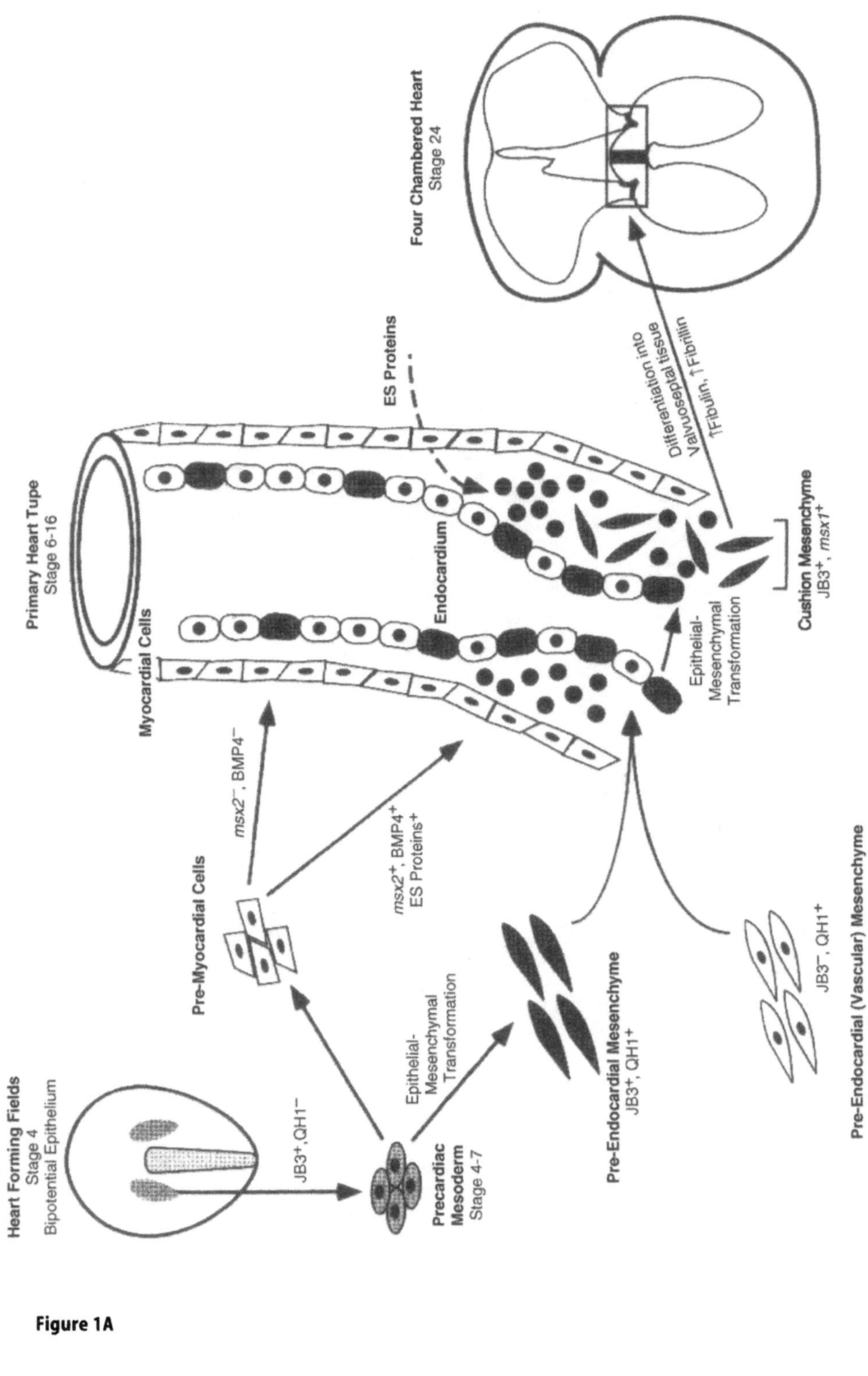
Heart Forming Fields
Stage 4
Bipotential Epithelium
JB3+, QH1−
Precardiac Mesoderm
Stage 4-7
Epithelial-Mesenchymal Transformation
Pre-Myocardial Cells
msx2−, BMP4−
msx2+, BMP4+
ES Proteins+
Pre-Endocardial Mesenchyme
JB3+, QH1+
JB3−, QH1+
Pre-Endocardial (Vascular) Mesenchyme
from Non-Cardiogenc Mesoderm
Myocardial Cells
Primary Heart Tupe
Stage 6-16
Endocardium
ES Proteins
Epithelial-Mesenchymal Transformation
Cushion Mesenchyme
JB3+, msx1+
Differentiation into
Valvuoseptal tissue
↑Fibulin, ↑Fibrillin
Four Chambered Heart
Stage 24

Figure 1A

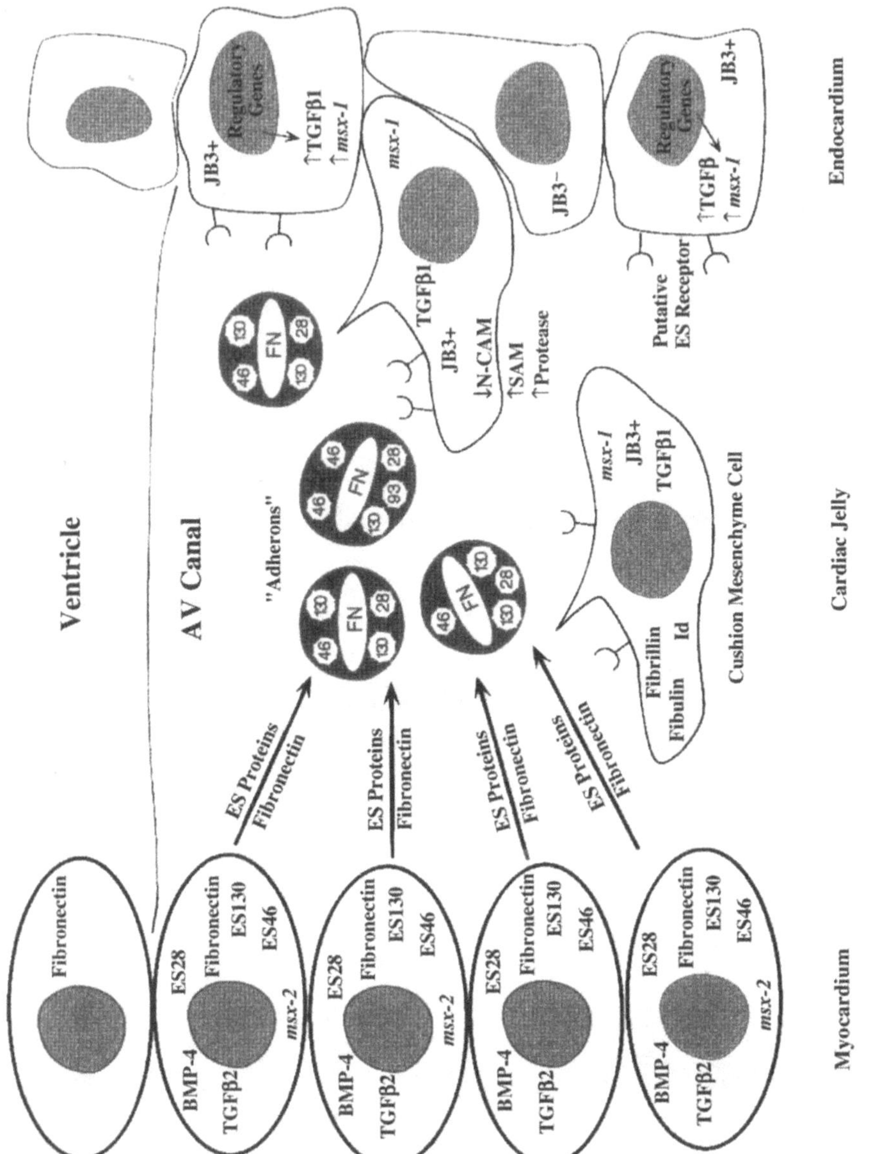
Ventricle
AV Canal
"Adherons"
Fibronectin
BMP-4
TGFβ2
ES28
ES130
ES46
msx-2
ES Proteins
Fibronectin
FN
46
130
28
93
Cushion Mesenchyme Cell
Fibrillin
Fibulin
Id
msx-1
JB3+
TGFβ1
↓N-CAM
↑SAM
↑Protease
Putative
ES Receptor
JB3+
Regulatory Genes
↑TGFβ1
↑msx-1
JB3−
↑TGFβ
Myocardium
Cardiac Jelly
Endocardium

Figure 1B

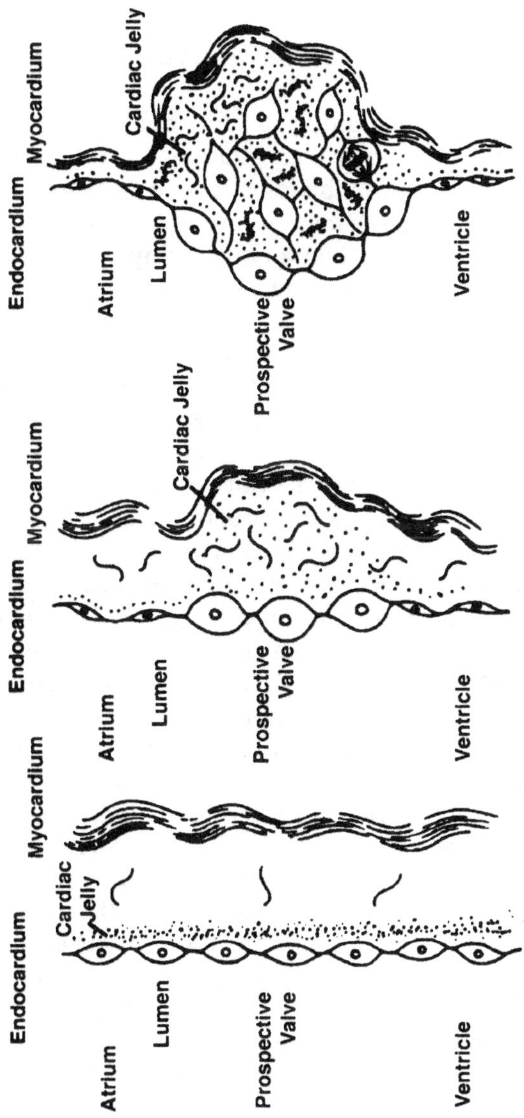
Endocardium
Myocardium
Cardiac Jelly
Atrium
Lumen
Prospective Valve
Ventricle
Endocardium
Myocardium
Cardiac Jelly
Atrium
Lumen
Prospective Valve
Ventricle
Endocardium
Myocardium
Cardiac Jelly
Atrium
Lumen
Prospective Valve
Ventricle

Figure 1C

4 Heart Valve Structure–Function Correlations: Valvular ECM and Cells

Normal cardiac valves permit unidirectional flow of blood without causing any of the following: (1) obstruction or regurgitation, (2) trauma to molecular or formed blood elements, (3) thromboembolism, or (4) excessive mechanical stress in the cusps and leaflets. Normal heart valve function requires structural integrity and coordinated interactions among multiple anatomic components in and around the valves. For the semilunar valves (aortic and pulmonary), the key structures are the cusps, commissures, and their respective supporting structures in the aortic and pulmonary roots. For the atrioventricular valves (mitral and tricuspid), the critical elements are the leaflets, commissures, annulus, chordae tendineae (tendinous cords), papillary muscles, and the atrial and ventricular myocardia.

The structure of the aortic valve (the valve most extensively studied, most frequently diseased, and most widely transplanted) best illustrates the essential concepts and will therefore be described in detail (Fig. 2). Aortic valve cusps open to form a nearly circular orifice during ventricular systole; they close rapidly and completely under minimal reverse pressure, stretching to maintain full competency throughout diastole (see Fig. 2A). The thickness of aortic valve cusps varies from approximately 300 to 700 μm throughout the cardiac cycle. The three aortic valve cusps (left, right and non-coronary) attach to the aortic wall at special areas of thickening known as the commissures. Behind the valve cusps are dilated pockets of aortic root, called the sinuses of Valsalva, from which the right and left coronary arteries arise from individual orifices behind the right and left cusps, respectively. Despite the pressure differential across the closed valve, which imposes a large load on the cusps, cuspal prolapse is prevented by substantial coaptation of the cusps so that they effectively support each other. The surface of apposition of the cusps in the closed phase of the valve is approximately 40% of the cuspal area; the coaptation area is a crescent-shaped portion of the cusp, termed the lunula. The pulmonary valve cusps and surrounding tissues are architecturally similar to the aortic valve, but the structures are more delicate than the corresponding aortic components, presumably due to the lower pressure environment in which the pulmonary valve functions.

All cardiac valves have a microscopically inhomogeneous architecture, consisting of well-defined, cellular tissue layers (see Fig. 2B) [23]. Similar to all the blood-contacting surfaces of the cardiovascular system, the valves are lined by a confluent layer of *valvular endothelial cells* (VEC); the cells deep to the surface are *valvular interstitial cells* (VIC) (see Fig. 2C). The principal components of valvular ECM are collagen, elastin and glycosoaminglycans. Adult human aortic cusps have approximately 43–60% dry weight of collagen, which is predominantly Type I (74%) mixed with Type III (24%) and Type V (2%), 10–13% elastin, and 20% GAGs [24]. Structural interrelationships among these components are critical to valve function [25]. Each of the three well-defined valvular layers is enriched in a specific ECM component. For the aortic valve, the thin

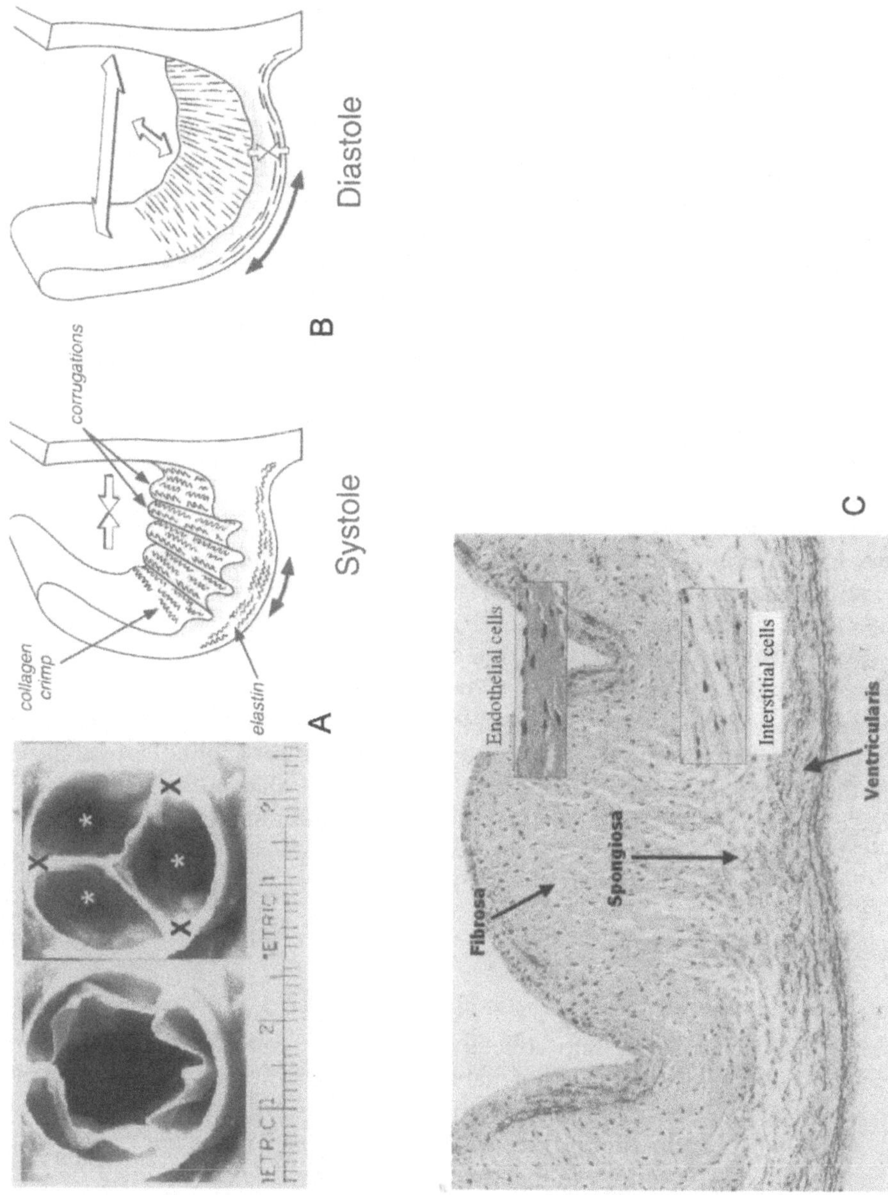

Fig. 2A–C Normal aortic valve structure and function. **A** In systole, the aortic valve cusps open to form a nearly circular orifice. They close rapidly and completely, stretching to maintain full competency throughout diastole. *X* Commissures, * sinuses of Valsalva. **B** Aortic valve cuspal architecture and configuration of collagen and elastin in systole and diastole. **C** Aortic valve histology. Low-magnification photomicrograph of cuspal cross-section, showing the three major layers and cuspal configuration in the non-distended state. Each of the black spots deep to the surface is the nucleus of an interstitial cell. The outflow surface is at *top*. Movat pentachrome stain. **A, C** reproduced by permission from [127], **B** reproduced by permission from [23]

layer closest to the left ventricular chamber (*ventricularis*) is comprised predominantly of collagenous fibers with radially aligned elastic fibers. The elastin of the ventricularis layer enables the cusps to have minimal surface area when the valve is open (left ventricular *systole*) but stretch in response to the back-pressure of the closed phase (*diastole*) to form a large coaptation area before recoiling in systole. The central layer of the aortic valve leaflet *(spongiosa)* is composed of loosely arranged collagen and abundant glycosoamino glycan. This layer has negligible structural strength, but accommodates the shape changes of the cusp during the cardiac cycle, lubricates relative movement between ventricularis and fibrosa layers and absorbs shock during closure. The *fibrosa* is a thick fibrous layer, which is composed predominantly of circumferentially aligned, densely packed collagen fibers, largely arranged parallel to the cuspal free edge and provides strength and stiffness. As normal valve cusps and leaflets are generally sufficiently thin to be perfused from the surrounding blood, normal human valves have few if any blood vessels, except at their points of proximal attachment to the vascular wall.

The orientation of architectural elements in the aortic valve cusps is anisotropic (i.e., nonrandom) in the plane of the tissue. This yields unequal mechanical properties of the valve cusps in different directions, with higher compliance in the radial than in the circumferential direction. Consequently, the geometry of the whole valve and the fibrous network within the cusps effectively transfers the stresses induced by diastolic back-pressure to the annulus and aortic wall. Additional structural specializations include: (1) folding (crimp) of collagen fibers along their length, and (2) orientation of bundles of collagen in the fibrous layer toward the commissures. This collagen orientation minimizes sagging of the cusp centers, preserves maximum coaptation, and thus prevents regurgitation.

4.1 Valvular Endothelial Cells

Endothelial cells (ECs) comprise the single-cell thick, continuous lining of the entire cardiovascular system, collectively called the endothelium. ECs are important contributors to the maintenance of vessel wall homeostasis and normal circulatory function. Uniquely containing *Weibel-Palade bodies,* 0.1 µm wide, 3 µm long membrane-bound storage organelles that contain von Willebrand's factor (vWF), P-selectin, and several other proteins, ECs can be identified immunohistochemically with antibodies to vWF and PECAM-1 (CD31).

With many synthetic and metabolic properties, ECs are active participants in blood–tissue interaction. They play a role in the maintenance of a nonthrombogenic blood-tissue interface, in the regulation of immune and inflammatory reactions, and in the regulation of growth of other cell types. Endothelium is a semipermeable membrane that controls the transfer of small and large molecules across the vascular wall. In most regions the intercellular junctions between ECs are normally impermeable to large molecules such as plasma proteins; however, these junctions are relatively labile and may widen when exposed to abnormal hemodynamic forces or vasoactive agents (e.g., histamine in inflammation).

Structurally intact ECs can respond to various pathophysiologic stimuli by adjusting their usual (*constitutive*) functions and by expressing newly acquired (*inducible*) properties, a process of phenotypic alteration termed *endothelial activation*. Inducers of endothelial activation include cytokines and hemodynamic forces. Activated endothelial cells produce adhesion molecules, other cytokines and chemokines, growth factors, vasoactive molecules that result either in vasoconstriction or in vasodilation, major histocompatibility complex (MHC) molecules, procoagulant and anticoagulant moieties, and a variety of other biologically active products. Normal endothelial function is characterized by a balance of these factors and the ability of the vessel to respond appropriately to various stimuli. In contrast, *endothelial dysfunction* (a form of endothelial activation) induces a surface that is thrombogenic or adhesive to inflammatory cells.

ECs have substantial phenotypic variability based on anatomic site and dynamic adaptation to local environmental cues. For example, EC populations that develop embryologically from different sites (large vessels vs capillaries, arterial vs venous) may have different characteristics [26]. We have recently shown that the endothelium from human pulmonary and aortic valves is phenotypically different [27], and that ECs of valves, which experience an altered mechanical environment, adapt to the new conditions. Moreover, several lines of evidence suggest that the valvular endothelial cells (VEC) are different than ECs from other sites. For example, a recent study has suggested that the endothelial response to flow is different on the valves than in the aorta [28], other emerging data suggest that there may be important phenotypic differences between the endothelium on the inflow and outflow surfaces of the aortic valve [135].

Defects in valvular endothelium and valvulogenesis have been noted in a number of gene "knockout" experiments in mice as was mentioned above. Studies from Joyce Bischoff's research group, in which we collaborated, showed that NFATc1 is required for maximal VEGF-induced proliferation in human adult pulmonary valve ECs in vitro [29]. Despite the lack of reported evidence for NFATc1 expression in post-natal murine valves, we also detected focal regions of NFATc1-positive endothelial cells along the leaflet surface. This finding suggests that the signaling pathway, which is active as endothelial cells proliferate during development, may be available in mature valves to potentially replenish endothelial and potentially interstitial cells as needed throughout adult life. Ovine aortic valve, but not vascular EC, can be induced to undergo an endothelial-to-mesenchymal transdifferentiation in vitro that is reminiscent of events that occur during valve development (see earlier).

Other evidence for differences in vascular and valvular endothelial responses comes from heart transplantation, where parenchymal leukocyte infiltration can result in organ rejection. Since vascularity is sparse in the cusps, inflammatory cells in the valves would have to be recruited from the circulation through the endothelium covering the valve cusp surface. However, histopathological studies of rejected orthotopic heart transplants show that their valves are spared the inflammatory cell infiltration and tissue damage that occurs in myocardium and is mediated through the microvasculature endothelium [30]. Valve cusps from pig hearts transplanted into baboons were also protected from leukocyte infiltration [31]. Since VCAM-1, ICAM-1 and E-selectin expression has been demonstrated in surgically removed diseased heart valves [32, 33] and in vitro, [34] the lack of vigorous inflammatory response in valve cusps from rejected hearts in cardiac transplant recipients cannot be explained totally by the absence or attenuated expression of these adhesion molecules. This suggests that other factors such as the hemodynamic forces in the vicinity of valves may contribute to the apparent protection from inflammatory cell infiltration. Understanding the function of VEC adhesion molecules in vivo in the context of a complex blood flow pattern, and indeed endothelial heterogeneity in general, requires further investigation.

4.2 Valvular Interstitial Cells

Our studies using normal, pathological and substitute valves have demonstrated that (1) valvular ECM is the principal determinant of valve durability, and (2) the quantity and quality of valvular ECM depend on viability and function of valvular interstitial cells (VIC), the most numerous cell type in the valves (recall Fig. 2C). VIC comprise a dynamic population of resident cells of multiple sources and phenotypes that synthesize the several types of valvular ECM molecules and express matrix degrading enzymes such as matrix metalloproteinases (MMPs), and their inhibitors (TIMPs), that mediate the ongoing ECM remodeling and repair. Studies over the past several decades have indicated that there exist diverse and dynamic VIC phenotypes, in a spectrum that ranges largely from fibroblast-like to myofibroblasts [35–37, 52, 58, 59]. Some investigators have suggested that contractile VIC may have a function in generating a cyclical valvular force that enhances cardiac function or aids adaptation to altered hemodynamic forces. In further support of the concept of functional VIC contraction, VIC proteins also include α and β myosin heavy chain and troponin isoforms [38], and valve leaflet contraction has been demonstrated in response to vasoactive agents [39]. Vasoactive agents also induce VIC growth and collagen synthesis [40].

We and others have evaluated valvular interstitial cell phenotype in normal and diseased human valves using immunohistochemistry for myofibroblast markers (α-smooth muscle actin, vimentin, and desmin) [27, 41, 42, 52, 58], and smooth muscle cell-specific myosin heavy chain isoforms (SM1 and SM2) [43]. These antibodies serve as useful markers to distinguish fibroblast-like and smooth-muscle cells from myofibroblasts. In normal mitral valves, interstitial cells (in situ) express vimentin, but not α-SMA, desmin or SM1 suggesting that those cells are fibroblast-like. Similarly, nearly all VIC in healthy adult human and sheep pulmonary valves are fibroblast-like cells immunoreactive to vi-

Table 2 Cellular density in heart valves (cells/mm^2)

	Pulmonary	Aortic
Human		
Fetal (n=18)		
14–19 week gestation	3032±190[a]	3032±190[a]
20–39 week gestation	2629±214	
Adult (n=5)	351±41	307±39
Adult by layer		
Fibrosa	378±63	325±36
Spongiosa	295±14	254±16
Ventricularis	380±14	350±68
Sheep (n=3)		
Neonatal	2563±145	
Juvenile (2 years old)	2130±127	2410±126

[a] Semilunar valves under equivalent pressures; specific location not specified.

mentin, but not α-SMA, SMemb, or MMP-13 ["normal/quiescent" phenotype: V+/A-/S-/M-, with 2.5±0.4% (human) and 4.9±0.7% (sheep) of cells α-SMA-positive]. In contrast, previous in vitro studies using isolated myofibroblasts from heart valves in culture demonstrated that 57–80% of cells are α-SMA-positive [36]. The higher percentage of myofibroblasts observed upon removal of cells from the intact valve and/or culture suggests that manipulation stimulates interstitial cell activation. Moreover, treatment of isolated valvular interstitial cells with TGF-β strongly activates interstitial cells to the myofibroblast phenotype (recall that TGF-β promotes endothelial to mesenchymal transdifferentiation) [136].

We have examined cellular density and distribution in different layers of aortic (NAV) and pulmonary (NPV) valves. Cellular density was measured as mean number of cells per high power field (cells in ten fields were counted and averaged), and expressed as cell number per millimeter squared of tissue section. No significant difference in cell number was noted between and in different layers (Table 2) of normal NAV and NPV. These data suggest that the cellular density is highest in fetal maturation and therefore decreases, perhaps continuously throughout life. This is a subject of our current research.

5 Valvular Adaptation, Repair and Regeneration

Valvular injury induces migration, proliferation and apoptosis of interstitial cells. Early events in valve repair can be studied in an in vitro model of a linear superficial denuding wound [44, 63]. In this model interstitial cells at the wound edge show prominent migration and proliferation. Interestingly, in response to a complete endocardial endothelial-denuding injury of the mitral valve, the interstitial cells from ventricular side of the valve proliferate at a higher rate than the interstitial cells from the atrial side of the valve [45]. VIC migration is likely regulated by the sequence of cellular processes including activation of integrins, cell surface heterodimeric receptors that mediate cell–cell and cell–ECM adhesion [46, 47] and other membrane receptors to provide cells with information about their external environment. Signal transduction mechanisms activate downstream pathways that regulate the distribution and activity of the cytoskeleton and associated proteins that lead to cell spreading, contraction and translocation. These events have been reviewed recently [48].

Our studies show that VIC become activated and mediate functional biomechanical adaptation of valves when they are exposed to a new, altered mechanical environment [27, 49, 50, 52, 58]. This activation is potentially reversible and when equilibrium is achieved, the cells again become quiescent. We hypothesize that VIC phenotype is regulated predominantly by the wall stresses in the valve leaflets, analogous to remodeling in cardiac hypertrophy and vascular walls. In leaflets from patients with myxomatous mitral valve degeneration, many cells expressed not only vimentin, but also α-SMA, desmin and SMemb (but not SM1) [58]. That the interstitial cells of these myxomatous valves express features of activated myofibroblasts, suggests that equilibrium is not achieved in valves with this condition. We also demonstrated that the immunophenotype of the cells in tissue-engineered valves progressed from activated myofibroblasts in vitro to fibroblast-like cells, and ultimately, to a cell phenotype characteristic of native valves in vivo [50]. This suggests that normalization of tissue stresses had been achieved. These cell phenotype changes resembled evolution of physiological wound healing in mitral valves associated with the phenotypic modulation of interstitial valvular cells from fibroblasts to myofibroblasts [51].

Replacement of the diseased aortic valve by transplantation of the patient's own pulmonary valve into the aortic position is performed in pediatric and young adult patients as an autologous tissue valve replacement (Ross procedure). To study cellular phenotypic changes and ECM remodeling of valves in response to altered mechanical loading in vivo, we analyzed the structure and interstitial cell phenotypes in ten clinical pulmonary-to-aortic valve transplants early (2–10 weeks) and late (3–6 years) postoperatively [27]. At all intervals, autografts showed near-normal cuspal structure and collagen architecture, and viable VIC. In early autograft cusp explants, 19.3±2.4% of VIC resembled myofibroblasts with α-SMA+/vimentin+ phenotype, and strong expression of MMP-13, indicative of ECM remodeling (see later). In contrast, only 6.0±1.1% of VIC in late explants and 4.6±0.8% of VIC in normal valves were α-SMA+/vimentin+. This study showed that pulmonary valves transplanted to the systemic circulation had early and reversible phenotypic modulation of VIC to myofibroblasts that secrete proteolytic enzymes. Subsequently, there was late normalization of VIC phenotypes and valve morphology toward those of the normal aortic valve [27].

Collectively, these data on dynamic VIC phenotype in several circumstances, including human and ovine fetal valves, mitral valves with myxomatous degeneration, short-term (ST) and long-term (LT) pulmonary autografts, and tissue-engineered valves in vitro and in vivo suggest a general mechanistic paradigm by which cardiac valvular tissue is dynamically and reversibly responsive to environmental conditions, particularly mechanical loading (Fig. 3) [52]. Under equilibrium conditions, the majority of the interstitial cell population in normal cardiac valves is composed of quiescent fibroblast-like cells. When stimulated by mechanical loading and other environmental stimuli (as in valve development, adaptation, pathology and substitution), VIC become activated and mediate connective tissue remodeling to restore the normal stress profile in the tissue. When equilibrium is restored, the cells return to the quiescent state.

VIC–matrix interactions are likely to be very important in regulating interstitial cell functions, including secretion of ECM components and proteolytic enzymes and cell migration, an important feature of all types of tissue repair. MMPs comprise a family of enzymes including interstitial collagenases and

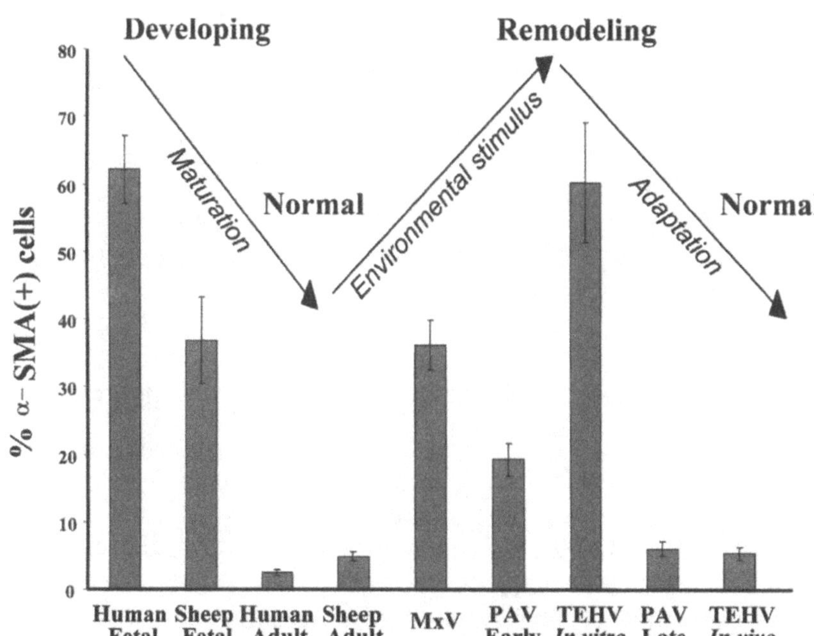

Fig. 3 Valvular interstitial cell plasticity demonstrated as dynamic and reversible through changes in phenotype shown as fraction of α-smooth muscle actin (*SMA*) positive cells in studied valve models. In response to different environmental stimuli (altered mechanical loading, injury) interstitial valvular cells during development, adaptation, disease and in tissue-engineered valves undergo activation and phenotypic modulation shown as fraction of α-SMA-positive cells. *MxV*, valves with myxomatous degeneration, *PAV*, pulmonary (to aortic) autograft valves; *TEHV*, tissue-engineered heart valves. Reproduced by permission from [52]

gelatinases involved in the degradation and remodeling of connective tissue [53]. MMPs play a pivotal role in normal tissue remodeling processes, such as tissue morphogenesis and wound healing. The interplay of MMPs, TIMPs and their regulators are especially important in cardiac and vascular remodeling. Interstitial collagenases (MMP-1 and MMP-13) mediate the initial step of collagen degradation by breakdown of the native helix of the fibrillar collagen network. These fragments then become accessible to the other proteases, such as gelatinases (MMP-2 and MMP-9), which further catabolize collagen [54]. In atheroma, cysteine endoproteases (cathepsin S and K) are involved in remodeling of ECM, particularly elastin [55]. Cathepsin K is the most potent elastase yet described, and also possesses collagenolytic activity. Several studies have suggested that ECM degradation in various degenerative diseases may be mediated by the cells residing in those tissues. This evidence suggests that these cells are being stimulated in some way to produce and secrete soluble extracellular messengers, which in turn interact with indigenous cells, thereby inducing them to initiate matrix degradation. Cardiac catabolic factor, derived from porcine heart valves was found to stimulate collagen and proteoglycan degradation in vitro [56].

Normal valves have a specific pattern of expression of MMPs and TIMPs [57]. Excessive levels of proteolytic enzymes such as collagenase-1 (MMP-1) and collagenase-3 (MMP-13) or gelatinase-A (MMP-2) and gelatinase-B (MMP-9) and cysteine endoproteases (cathepsin S and K) elaborated by VIC may contribute to collagen and elastin degradation, resulting in weakness and deformation of leaflets of heart valves with myxomatous degeneration [58]. An important role of proteolytic enzymes in matrix remodeling in myxomatous valve disease as well in tissue-engineered heart valves and in clinical pulmonary autograft aortic valve substitutes seems likely.

6
Information from Studies of Contemporary Bioprosthetic Heart Valves

Structural dysfunction of tissue valves is the major cause of failure of the most widely used bioprostheses (flexible-stent-mounted, glutaraldehyde-preserved porcine aortic valves and bovine pericardial valves) [4, 60]. Within 15 years following implantation, 30–50% of porcine aortic valves implanted as either mitral or aortic valve replacements require replacement because of primary tissue failure [61]. Cuspal mineralization and non-calcific damage to the cuspal structural matrix are the major pathologic processes responsible. The most frequent failure mode is regurgitation through secondary tears [4, 62]. Pure stenosis, owing to calcific cuspal stiffening and noncalcific cuspal tears or perforations (reflecting direct mechanical destruction of collagen), occurs less frequently. Calcific deposits are usually localized to cuspal tissue (intrinsic calcification), but calcific deposits extrinsic to the cusps may occur in thrombi or endocarditic vegetations [4, 64]. Calcification is markedly accelerated in younger patients, with children and adolescents having an especially accelerated course.

Alterations in tissue structure and chemistry of bioprosthetic valves that occur during valve fabrication and manufacture can contribute to failure modes [49]. The most important changes and their consequences are as follows:

1. The microstructure of the cusps is fixed in a static geometry characteristic of one phase of the cardiac cycle; thus, the normal functional cyclical rearrangements in valve substructure cannot properly occur. This induces abnormal tissue stresses.
2. Damage to the endothelial coverage allows penetration of plasma and inflammatory cells into the cusp.
3. The interstitial cells are nonviable due to the cytotoxicity of chemical treatments used to cross-link collagen; without viable cells, synthesis of collagen or other ECM components cannot occur.

The principal processes that account for limited bioprosthetic valve durability in vivo after implantation are cuspal mineralization and noncalcific mechanical fatigue. The mechanism of calcification involves the reaction of calcium derived from plasma with organic phosphorous residing in the cross-linked, nonviable cells of the preserved valve [65–67]. In addition, proteolytic degradation of the collagenous ECM may occur [68] and mechanical damage could cause noncalcific ECM deterioration or result from mechanical fatigue effects. Damage owing to purely mechanical effects has been noted in explanted porcine aortic valves, and increased levels of ECM degrading activity, MMPs, including MMP-2 and MMP-9, have been found in explanted clinical bioprostheses [69]. The nature of the in vivo structural damage patterns noted in explanted bioprosthetic valves are similar to those found in vitro using accelerated durability testing, which suggests that pure mechanical effects are likely important in ECM degradation [68]. Calcific and noncalcific damage, both mechanical and chemical, could be synergistic.

7
Conceptual Approach to Heart Valve Tissue Engineering

The rationale and design criteria for tissue-engineered heart valves (TEHV) derive from concepts established over the past decade from studies in several laboratories of normal, developing and diseased heart valves, bioprosthetic valves, and other tissue valve substitutes (Table 3). Although these studies largely focused on the elucidation of the mechanisms responsible for specific clinical observations, they have indicated the potential of finding useful markers of cell function and matrix structure and physiology and have led to five key concepts of functionally adaptive valvular remodeling/regeneration:

1. The highly specialized arrangement of collagen and other ECM components (particularly elastin and proteoglycans) enables normal heart valve function and is the principal determinant of the durability of heart valves.

Table 3 Studies of heart valve pathology that have informed tissue engineering

Valve models	Application to TEHV
Normal native valve	– Initial cell and scaffold and seeding conditions – End points for TEHV remodeling in vitro and in vivo
Developing valve	– Mechanisms of TEHV cell-cell and cell-matrix interactions – Cellular response to different hemodynamic conditions
Diseased valve	– Valvular cell and ECM response to a range of environmental stimuli
Viable valve transplant (pulmonary autograft)	– Cellular response to altered environmental conditions
Glutaraldehyde-pretreated bioprotheses and cryopreserved allograft valves	– Importance of ECM structure and properties – Potential for complications encountered in current tissue valves

2. Structural deterioration of native and substitute valves is ultimately mediated by chemical and mechanical damage to collagen.
3. The quality of valvular ECM depends on valvular interstitial cell viability, function and ability to adapt to different environments.
4. Cell viability in nearly all current bioprosthetic tissue valve substitutes is compromised or completely eliminated during processing; thus, ECM damage, which occurs during valve function following implantation of current bioprosthetic valves, cannot be repaired.
5. The long-term success of a tissue-engineered (living) valve replacement will, therefore, depend on the ability of its living cellular components to assume normal function with the capacity to repair structural injury, remodel the ECM, and potentially grow.

The innovative fabrication of materials and the development of sophisticated methods to repair or regenerate damaged or diseased tissues and to create organ replacements requires integration of a diverse array of basic scientific principles and enabling technologies. Thus, the ability to engineer tissue requires an understanding of relationships of structure to function in normal and pathological tissues (including mechanisms of embryological development, and functional tissue biomechanics and other structure-function correlations), and the ability to control cell and tissue responses to injury, physical stimuli and biomaterial surfaces, through chemical, pharmacological, mechanical and genetic manipulation. The immense need for engineered cardiovascular tissues has generated considerable interest and investigation [70–73, 128].

While much work is currently underway to improve current valve bioprostheses, including design optimization and anticalcification pretreatments, ma-

Table 4 Expanded design criteria for a tissue-engineered heart valve

Functional as implanted
Endothelialized blood-contacting surface
Cellular potential for extracellular matrix synthesis, remodeling, and repair
Appropriate heterogeneity, anisotropy, and amount of extracellular matrix
Potential growth with patient
Absence of deleterious immunological and other inflammatory processes
Resistance to calcification
Resistance to tissue overgrowth
Stable geometry and mechanical properties
Large effective orifice area
Prompt and complete closure
Resistance to infection
Chemically inertness and lack of hemolysis
Easy and permanent insertion

Modified from [4].

jor impact will require radically new technologies. This is especially the case in pediatric applications, where physiologically corrective procedures can be successfully performed, but those involving valve replacement necessitate reoperations to accommodate growth of the patient. In addition, repairs of congenital deformities may require very small valve sizes that are not commercially available. The use of engineered tissue offers the potential to create a non-obstructive, non-thrombogenic tissue valve substitute containing living cells capable of providing ongoing remodeling and repair of cumulative injury to the ECM and allowing growth in maturing recipients. The design criteria for an engineered living valve, capable of repair, remodeling and adaptation, are summarized in Table 4.

Tissue engineering is a broad term describing a set of tools at the interface of the biomedical and engineering sciences intended to produce therapeutic or diagnostic medical devices that have living cells or attract endogenous cells to aid tissue regeneration [74]. In the general paradigm of tissue engineering (Fig. 4), cells are seeded on a synthetic polymer or natural material which serves as a *scaffold* and then a tissue is matured in vitro (*bioreactor*), until proliferating cells produce a sufficient amount and quality of ECM to form the *construct*. In the second step, the construct is implanted in the appropriate anatomic location, where further remodeling in vivo may occur to recapitulate the normal functional architecture of an organ or tissue. The key processes occurring during the in vitro and in vivo phases of tissue formation and maturation are (1) cell proliferation, sorting and differentiation, (2) ECM production and organization, (3) degradation of the scaffold, and (4) remodeling and potentially, (5) growth of the tissue.

Some cardiovascular tissue engineering approaches can employ "*incomplete*" paradigms (Fig. 5), which omit a step or steps in the general paradigm by taking advantage of specific properties of initial components (cell or scaffold) and characteristics of engineered tissue. For example, the *cell-seeded scaffold* model

A

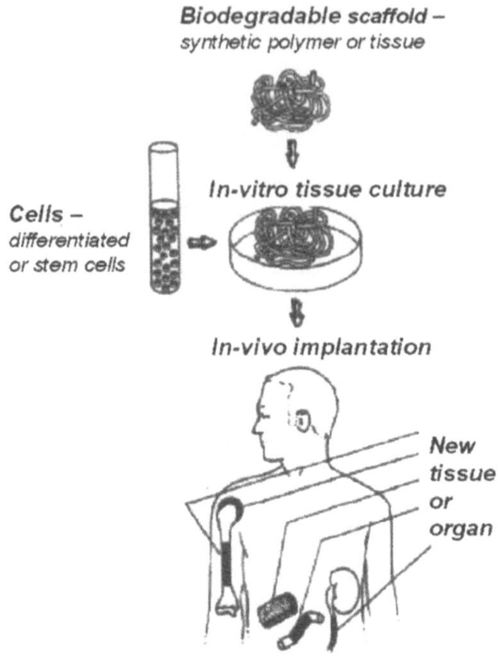

B

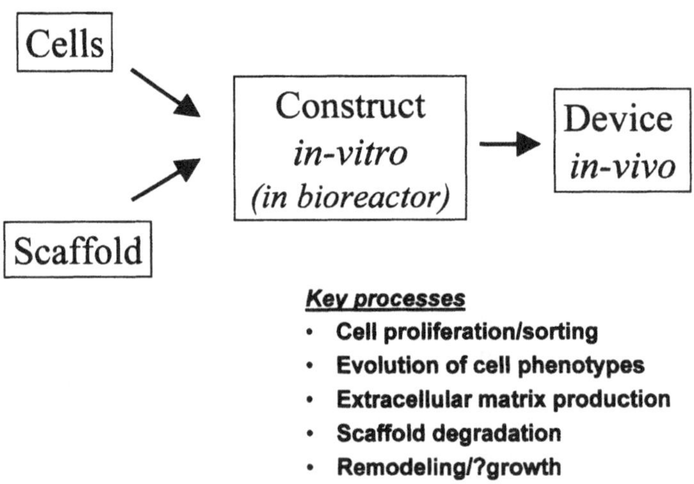

Fig. 4A, B Tissue engineering paradigm. **A** The tissue engineering process. **B** In the first step of traditional tissue engineering approach, differentiated or undifferentiated *cells* of different sources are seeded on a bioresorbable *scaffold* and then the *construct* matured in vitro in a bioreactor until the cells proliferate and elaborate extracellular matrix to form a "new" tissue. In the second step, the construct is implanted in the appropriate anatomical position, where remodeling in vivo is intended to recapitulate the normal tissue/organ. Several key pathophysiological processes are occurring during the in vitro and in vivo phases. **A** Modified and reproduced by permission from [128]

offers the possibility of putting a confluent endothelial coverage on a vascular graft at the time of implantation, and therefore the first step (in vitro tissue maturation) is omitted (see Fig. 5A) [75, 76]. The *cell transplant* model used in experimental and early clinical trials of cell-based therapy for myocardial tissue repair uses injected cells at a specific tissue site, in the hope that these cells will differentiate and proliferate, and/or stimulate repair by functional tissue (see Fig. 5B) [77, 78]. More recent approaches to valve and vascular graft engineering have utilized *scaffolds that attract endogenous cells* to repopulate and remodel a decellularized matrix (see Fig. 5C) [79, 80]. Cardiovascular tissue engineering may also make it possible to design physiological in vitro *tissue models* to study cell function and the pathogenesis of cardiovascular diseases, and ultimately allow molecular treatment of diseases that are currently considered potential targets for tissue engineering (see Fig. 5D) [81–83].

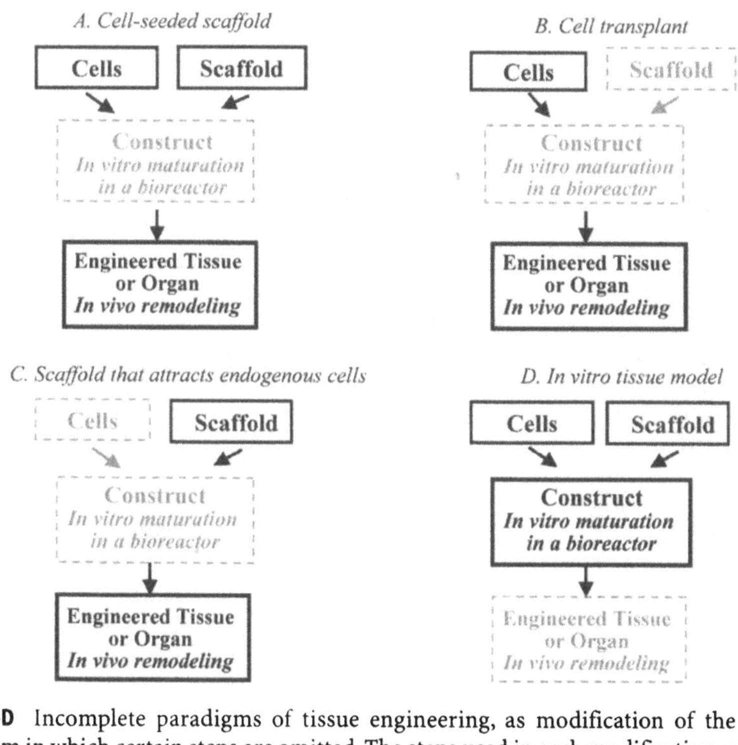

Fig. 5A–D Incomplete paradigms of tissue engineering, as modification of the general paradigm in which certain steps are omitted. The steps used in each modification are shown in solid black lines and text; omitted steps have dotted/light lines and text. **A** *Cell-seeded scaffold* model exemplified by endothelial cell seeding of vascular graft at the time of implantation. **B** The *cell transplant* model has been used to approach cellular cardiomyoplasty in myocardial tissue engineering. **C** Valve and vascular graft tissue engineering utilizing, for example, decellularized *scaffolds that attract endogenous cells* to repopulate and remodel an altered tissue. **D** Physiological in vitro *tissue models* will help to understand physiological processes, pathogenesis of diseases, and provide diagnostic and therapeutic tools. Reproduced by permission from [74]

8 Control of Structure and Function in Engineered Tissues

Biological and engineering challenges in tissue engineering have been focused on the three principal components that comprise the "cell-scaffold-bioreactor" system (see Fig. 4B). Control of the various parameters in device fabrication may have a major impact on the ultimate result (Table 5).

Table 5 Control of structure and function in tissue-engineered heart valves

Cells
Tissue source, age, species
Cell type, heterogeneity within cell type
Phenotype (differentiated vs. stem cell, modification, dysfunction)
Preseed culture conditions
Viability
Method of preservation
Mechanotransduction
ECM synthesis and remodeling
Scaffold
Chemistry
Configuration/porosity
Heterogeneity (architecture, composition)
Cell adherence
Biocompatibility
Incorporation of bioactive molecules
Degradation rate (in vitro, in vivo)
Mechanical properties (strength, compliance)
Ease of manufacture
Biological signals
Source
Pure vs tissue extract
Single vs multiple
Organ specificity
Function (effect on adhesion, migration, proliferation, and sorting)
Drug delivery
Genetic manipulation
Construct
Bioreactor medium
Static vs flow
Growth factors, morphogenetic controls
Degree of maturation at implantation
Functionality
Biocompatibility
Sterilization, storage
Quality control

From [4].

8.1
Cell Sources

Ideally, the cells used to seed the developing tissues should be nonimmunogenic, capable of expansion in vitro to yield increased numbers of cells, easy to harvest, and have the ability to provide specific cell functions. Theoretically these cells may be of allogenic, xenogenic, or autologous source. Limitations intrinsic to donor allogenic and xenogenic cells include immunorejection or the possibility of virus/pathogen transmission [84]. However, autologous cells obtained from the patient also have limitations which include: (1) sparse availability of healthy tissue, (2) the time required for cell expansion affecting patient survival and product cost, (3) autologous cells not being amenable to off-the-shelf availability, (4) structures having possible product-to-product variability, and (5) cells from older individuals possibly having diminished proliferative ability. Cells can be of different sources: (1) differentiated cells from primary tissues (e.g., valvular cells or cardiomyocytes), (2) tissue-specific adult stem cells (e.g., hematopoietic or neural stem cells), (3) bone marrow stromal cells (e.g., mesenchymal progenitor cells) [85], (4) bone marrow-derived, circulating stem cells including endothelial and smooth muscle cell precursors [86], and (5) pluripotent embryonic stem cells. The potential use of endogenous or exogenous multipotential, mesenchymal stem cells for seeding of valves and other cardiovascular devices is particularly exciting, and may obviate the issue of a pre-

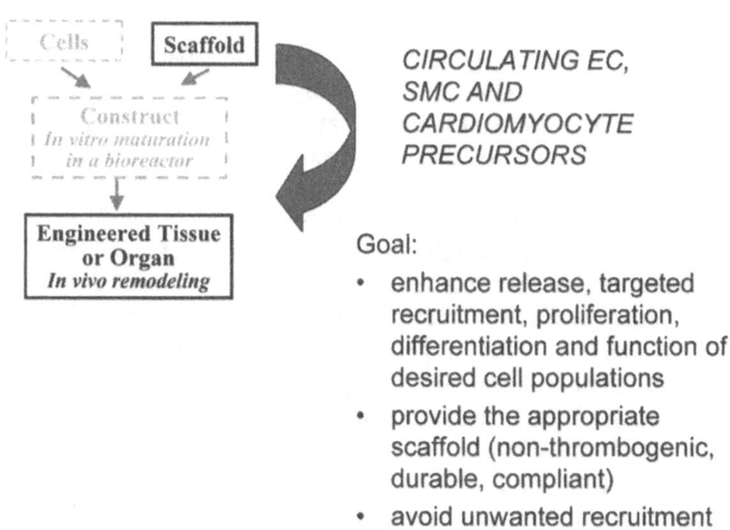

Fig. 6 Potential role of circulating endothelial (*EC*) and smooth muscle cells (*SMC*) in tissue engineering. Studies indicate circulating cells that are capable of contributing to valve, vascular wall and cardiac healing. The challenge is to understand and control the recruitment, differentiation, proliferation and function of these cells in order to maximize their role in tissue regeneration

selected cell phenotype [87, 88]. However, ethical concerns that relate largely to specific products that use human embryonic stem cells [89], human embryonic and fetal tissue [90], and xenotransplantation [91] are still debated and this could inhibit their clinical application.

Cell sources for engineered heart valves have included human and animal arterial wall cells, mesenchymal and bone-marrow-derived stem cells, or valvular cells obtained by experimental biopsy. It is of particular interest in this context that: (1) transdifferentiation of endothelial cells to interstitial cells occurs during embryonic valve development and possibly postnatally (discussed earlier), and 2) there is considerable evidence that a population of circulating endothelial and host bone-marrow-derived endothelial and, smooth muscle precursor cells likely have the ability to seed a blood-contacting surface and potentially populate a valve [92–95] (Fig. 6). However, the quantitative extent to which any of these recently discovered endogenous cell types can contribute to functional cells is yet unknown.

8.2 Biomaterial Scaffolds

A scaffold material for use in tissue engineering must be (1) biocompatible, (2) bio-degradable into non-toxic products, and (3) manufacturable; additionally, it should have (4) a highly porous and yet mechanically stable macro-structure to facilitate cell ingrowth, and (5) a surface chemistry that induces cells to attach and regenerate complex tissues. In the future, the scaffolds may include bioactive molecules and/or ligands that induce specific and desired cell responses and regulate cellular gene expression and cellular organization ("biomimetic" scaffolds) (Fig. 7) [96, 97]. Scaffolds can be produced from natural or synthetic biomaterials. Natural materials are usually composed of ECM components (e.g., collagen, elastin, fibrin, or glycosamino-glycans or decellularized tissues, such as heart valve, pericardium, arterial wall or small intestinal submucosa) [98, 99]. Decellularized tissue has been used as a scaffold that is designed to either be seeded with living cells in vitro which can attract selected cell populations in vivo that will achieve the appropriate topographic distribution and differentiation of the natural cells in that location.

Synthetic polymers are advantageous in that their chemistry and material properties such as macro-structure (e.g. porosity, pore size, mechanical stability, degradation time) [100] and micro-structure (e.g. three-dimensional structure at a resolution of ~200 μm, microfabricated membranes with channels at a resolution of 40–310 μm) [101] may be well-controlled. The most widely used synthetic polymers in tissue engineering have included the poly α-hydroxyesters poly(glycolic acid) (PGA), poly(L-lactic acid) (PLLA), and copolymers poly(lactic-co-glycolic acid) (PLGA), and poly(anhydrides) and poly(peptides). Natural polymers such as polyhydroxyalkanoates (PHA) are thermoplastic, biocompatible, resorbable, flexible, and induce only a minimal inflammatory response [102]. Elastic biodegradable polymers are now available [103].

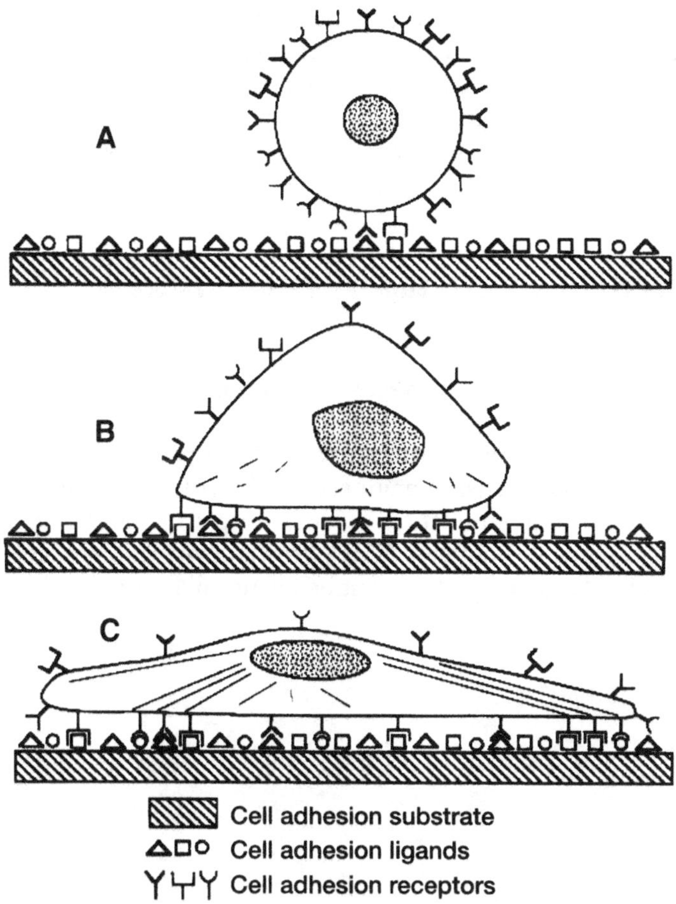

Fig. 7A–C Progression of receptor-ligand interaction dependent adhesion of a cell on a biomaterial scaffold. **A** Initial contact. **B** Bonding of cell surface receptors with surface-attached cell adhesion ligands. **C** Cytoskeletal reorganization with progressive spreading of the cell on the substrate. Reproduced by permission from [129]

8.3 Bioreactor

The bioreactor can provide three-dimensional tissue developing in vitro with appropriate biochemical mass transport (e.g., oxygen and growth factors) and mechanical stimulation (e.g., tension/compression and shear stress), mimicking conditions present in vivo. Early bioreactors were designed simply to pump nutrient liquid (culture medium) to the assembling tissue. The next generation of bioreactors subjected the tissue to tension, compression and shear stresses and pulsatile flow of culture medium. Bioreactor conditions have been shown to affect cell viability and proliferation, and ECM content and architecture in engineered tissue and thereby mechanical properties of engineered, mechanically responsive tissues such as valves, vessels, cardiac muscle, and cartilage [104]. As mechanical forces are also potent regulators of cell-mediated growth and degradation and repair of cardiovascular and musculoskeletal tissues, elucidation of the mechanisms involved in cellular signaling induced by mechanical forces (*mechanotransduction*) will be needed to develop these new strategies.

9 Current Status of Research in Engineered Heart Valves

9.1 Heart Valve Constructed from Biodegradable Polymer Scaffold

The heart valve tissue engineering group at Children's Hospital, Boston has fabricated and implanted tissue-engineered valve cusps and valved conduits in the pulmonary arterial circulation in lambs (Fig. 8) [105]. Using autologous cells and biodegradable polymers, TEHV grown in vitro have functioned in the pulmonary circulation of growing lambs for up to 5 months, and evolved to

A 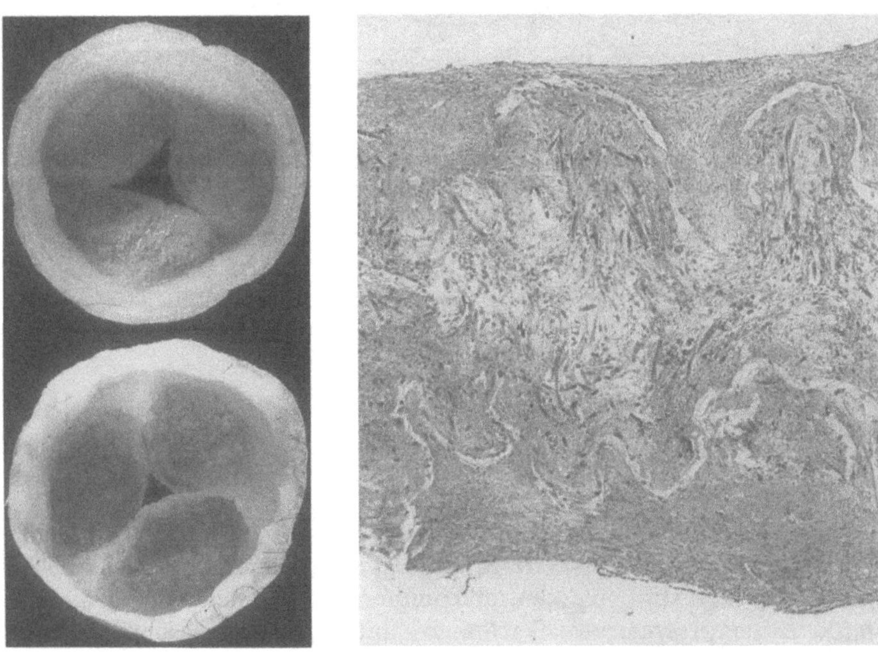 B

Fig. 8 A, B Tissue-engineered heart valve after 14 days of conditioning in bioreactor. **A** Gross morphology of engineered heart valve. **B** Histological section of leaflet portion of tissue-engineered valve after 14 days of pulsatile flow demonstrated dense fibrous tissue near surface and loose central core (hematoxylin and eosin, magnification ×200). **A** Modified and reproduced by permission from [105]

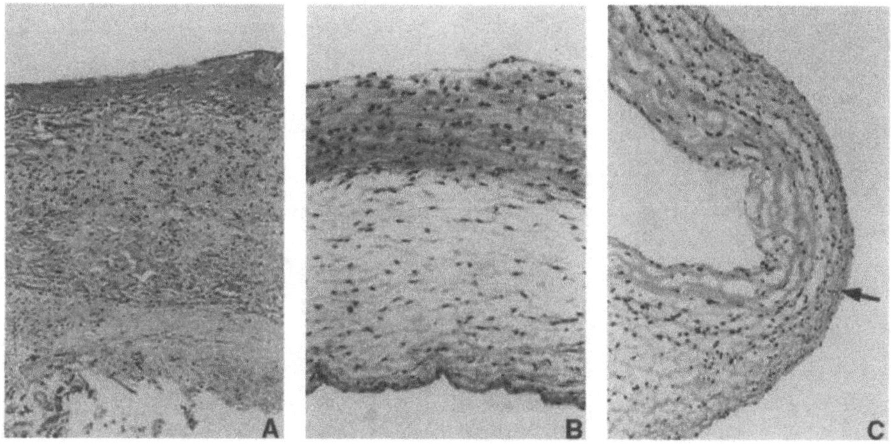

Fig. 9 A–C Histology of engineered heart valve leaflet in vivo. **A** At 6 weeks, there is early organization of tissue predominantly in outer layer (*top*) (magnification ×50). **B** Cross section of leaflet at 16 weeks shows layered cellular fibrous tissue, which is more dense near outflow surface (*top*) (magnification ×100). **C** Cross section of leaflet at 20 weeks demonstrates layered cellular fibrous tissue, which is more dense near outflow surface (*top*), with elastin (shown as *black streaks*, highlighted by *arrow*, near inflow surface; magnification ×100). **A** and **B**, hematoxylin and eosin stained; **C** stained with the Movat stain. Modified and reproduced by permission from [105]

a specialized layered structure that resembles that of native valve (Fig. 9). Trileaflet valve constructs were fabricated from poly-4-hydroxybutyrate-coated polyglycolic-acid, seeded with autologous ovine endothelial and carotid artery medial cells, and cultured in vitro up to 14 days in a bioreactor that mimicked valve cyclical motions by providing pulsed flow of cell culture medium. The in vitro cell culture system provided physiological pressure and flow of nutrient medium to the developing valve constructs in a dynamic cell culture setting. After 14 days in the bioreactor the ECM that was produced was predominantly proteoglycan (Fig. 10). Collagen accumulation was detected by picrosirius red staining observed under polarized light after 14 days in the bioreactor as a few fibrils of weak predominantly green birefringence (collagen type III). The collagen fibrils were oriented along the spindle-shaped cells detected throughout the tissue (Fig. 11). Elastin was not detected histologically. Immunohistochemical staining revealed that cells in the construct were activated myofibroblasts, as determined by strong expression of α-SMA (microfilaments), vimentin (intermediate filaments), SMemb (produced by activated mesenchymal cells), and MMP-13 co-expression.

The constructs were implanted as pulmonary valves in lambs: native pulmonary leaflets were resected, and 2-cm segments of the main pulmonary artery just distal to the native valve location were replaced by autologous cell-seeded heart valve constructs and evaluated at intervals up to 20 weeks. Additional

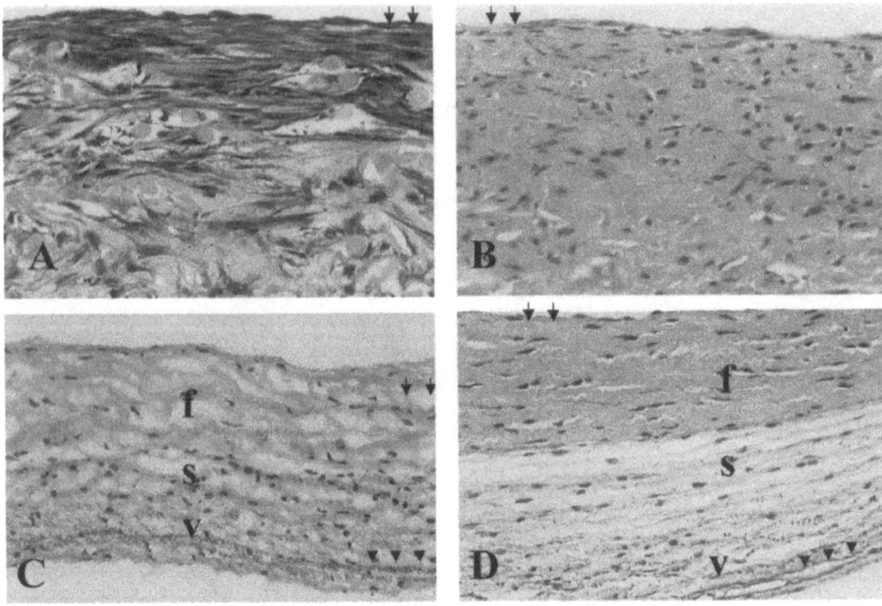

Fig. 10 A–D Evolution of tissue toward three-layered valve structure in tissue-engineered heart valves in vitro and in vivo. **A** Formation of ECM in vitro after 14 days in bioreactor and **B** in cusp at 4 weeks implants as predominant proteoglycan accumulation. Elastin was not detected. **C** Twenty weeks explant demonstrates three-layered structure: collagen was prevalent in the fibrosa (*f*), proteoglycans were dominant in the spongiosa (*s*), and elastin (stained as *thin, wavy black linear* material, highlighted by *arrowheads*) in the ventricularis (*v*), similar to native pulmonary valvesphown in **D**, Movat pentachrome stain. Outflow surface indicated by *arrows*. Circumferential sections of cusps; original magnification ×100. Modified and reproduced by permission from [50]

studies in vitro showed that the valves had minimal regurgitation, and only mild transvalvular gradients. Valves were explanted at 4–20 weeks and demonstrated the dynamic changes of interstitial cell phenotype and ECM in tissue-engineered valve explants toward layered architecture and cellular configuration of native valves. Some cells in explants at 4–8 weeks showed weak staining for α-actin and vimentin, but all cells throughout the leaflet stained strongly for SMemb and MMP-13. Explants at 16–20 weeks contained fibroblast-like cells, with predominant expression of vimentin and undetectable levels of α-SMA. Some cells were still positive for SMemb and MMP-13 (Fig. 12). Leaflets were partially covered with vWF-positive cells that were characteristic of endothelial cells. At 16–20 weeks, the leaflets were layered and had virtually uniform structure from base to edge. Movat stain demonstrated collagen in the fibrous layer on the aortic (outflow or sinus of Valsalva) side and proteoglycans in the central loose layer. Elastin was detected on the ventricular (inflow) side. Sections stained with picrosirius red and examined under polarized light further demonstrated three-layered collagenous architecture. Collagen fibers in

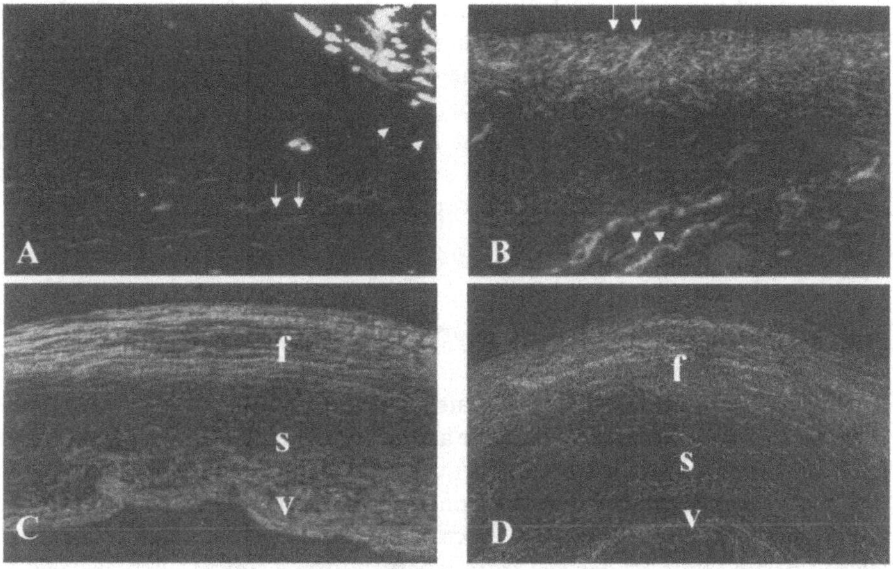

Fig. 11A–D Evolution of collagen in tissue-engineered heart valve. **A** In 14 days collagen was evident in in vitro construct as a few fibrils of weak birefringence (*arrows*); **B** 4 weeks implants show disorganized and disoriented accumulation of fibrillar collagen on the free edge of the leaflet (*arrows*). Polymer residue indicated by arrowheads in **A** and **B**. **C** Twenty weeks explant demonstrates three-layered collagenous architecture reminiscent of native valve **D**., (*f* fibrosa, *s* spongiosa, *v* ventricularis). Outflow surface indicated by *arrows*. Picrosirius red stain viewed under polarized light. Circumferential sections of cusps original magnification ×100. Modified and reproduced by permission from [50]

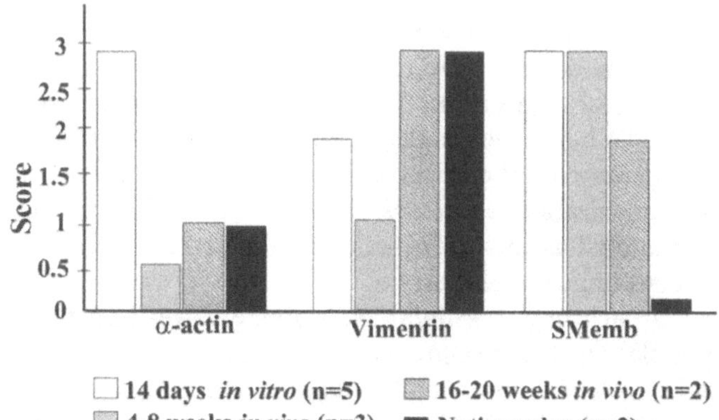

Fig. 12 Dynamic progression of cell phenotype in TEHV toward phenotype of interstitial cells of native valves: (1) regression of α-SMA expression, (2) progression of vimentin expression, (3) incomplete regression of SMemb marker of cell activation. Yet unknown is whether longer period of implantation will reduce cell activation entirely. Modified and reproduced by permission from [74]

the fibrous layer were thick, predominantly type I, densely packed, and arranged parallel to the free edge of the leaflet, reminiscent of native valve. Collagen was almost undetectable in the central layer, and appeared loose and fragmented below the ventricular side.

Thus, we have demonstrated that a tissue grown in vitro can function as a valve replacement in vivo and serve as a template for remodeling of tissue toward a structure with the morphologic characteristics of native valve cusps.

9.2 Heart Valves Fabricated From Naturally Derived Biomaterials

An alternative tissue engineering strategy uses non-aldehyde-preserved naturally derived biomaterials to fabricate a functional valve. This approach differs from conventional bioprosthetic heart valves in that the materials are designed to attract and provide a fertile environment for the adherence and growth of circulating endothelial and other precursor cells. These have included decellularized tissue scaffolds derived from a natural tissue source, such as valve or pericardium, or natural degradable polymeric scaffolds, such as collagen or fibrin gel. Natural valve scaffolds possess desirable three-dimensional architecture, mechanical properties, and potential adhesion/migration sites for cell ingrowth. Indeed, natural biomaterials are composed of ECM proteins that are conserved among different species and could serve as intrinsic templates for cell attachment, proliferation, and maintenance of cell viability and phenotype [106]. Model studies emphasize the potential of this approach. Freeze-dried porcine valves seeded in vitro with human endothelial cells and fibroblasts were reported to form a confluent endothelial layer on the surface of the leaflet, and viable fibroblasts were said to be present within the tissue [107]. Decellularized valves have been seeded with autologous or allogenic cells and implanted [108, 109]. These investigators report that valves showed normal function and complete histological restitution of valvular tissue with confluent endothelial surface coverage. Other results suggest that photooxidized bovine and porcine pericardium support endothelial cell growth in vitro [110]. However, the extent to which decellularization processes alter the valve ECM are not clear, and aneurysm formation has been reported in decellularized arterial wall grafts implanted in the systemic circulation.

Another approach uses decellularized porcine valves implanted without in vitro cell repopulation, thereby allowing and depending on endogenous ingrowth of host cells [111]. The commercially available SynerGraft valves prepared by CryoLife (Kennesaw, GA) exemplify this approach. Histological examination of such valves after up to 6 months in vitro function in sheep [112] and humans [113, 114] suggested intact leaflets had some in-growth of host cells and no evidence of calcification. However, degenerative structural failure of the SynerGraft heart valves has been reported in pediatric patients within 1 year [115]. Moreover, the matrix of SynerGraft valves elicited a strong inflammatory response that was non-specific early on and was followed by a lymphocyte response. There was no effective host cell repopulation of the valve matrix. Although detailed studies have yet to validate this general concept, it is an exciting possibility in that it may eliminate the problems associated with xenogenic cells, avoid calcification associated with glutaraldehyde fixation, and permit investigation of a biologic matrix that would provide a matrix microenvironment suitable for cellular repopulation.

Cell-free porcine small intestine submucosa has also been used as a resorbable bio-scaffold for site-specific tissue remodeling [116]. Normal swine had one pulmonary valve cusp excised, replaced with a leaflet constructed from a single layer of porcine small intestine submucosa, and explanted at 111 days postoperatively. Analysis of the explanted substituted valve cusps revealed resorption of the submucosal matrix, progressive replacement with fibrous connective tissue, progression of endothelialization and a layered architecture reminiscent of natural valve structure. This approach appeared to yield complete replacement of the implanted valve leaflet in contrast to the valve-based scaffold, discussed above, in which cell repopulation occurred with retention of the original ECM.

10 Challenges Associated with Engineered Heart Valves

The objective of creating a functional TEHV composed of collagenous and non-collagenous elements containing the endothelial and interstitial cells will be achieved if, and only if, the valve has an appropriate layered structure, and the cells are capable of renewing the complex ECM elements progressively degraded during function. In addition, no new complications should be introduced. This goal requires substantial advances in both scientific understanding and enabling technology, including development of techniques to non-invasively monitor the quality of tissue healing and remodeling in order to control the various parameters in device fabrication.

Tissue engineering represents a paradigm shift in heart valve development. A key issue relates to the need for acceptance by the surgical community. While the need is more compelling for the pediatric population, the use of a new valve with potentially complex problems in place of available valves with proven outcomes over the now-expected 15 or more years may be problematic [117]. Previous and existing heart valve substitutes were designed to behave as similarly as possible in all recipients; however, the response to engineered tissues may have marked patient-to-patient variability. This is conceptually analogous to the emerging area of pharmacogenomics [118] in that individual patient characteristics (for example, genetic mutations or polymorphisms in molecules important in matrix remodeling) could have a profound influence on outcome in some patients [119, 120]. Indeed, individuals with genetic defects in coagulation proteins may be unusually susceptible to thrombosis of prosthetic heart valves [121]. Thus, it will be increasingly important not only to understand how patient factors and response to injury control tissue structure and function but also to assess the progression of healing and remodeling in vivo in an individual patient.

In this regard, novel and innovative approaches must be used in the investigation of in vivo tissue compatibility. In the future, the assessment of bioactive and especially tissue-engineered implants will require a broadened scope of the concept of "biocompatibility" and the approaches employed in implant retrieval and evaluation will require identification of the phenotypes and functions of cells and the architecture and remodeling of ECM [122]. Thus, implant retrieval and analysis techniques will necessitate the identification of tissue characteristics (*biomarkers*) that will be predictive of (*surrogates for*) success and failure. A most exciting possibility is that such biomarkers may be used to non-invasively image/monitor the maturation/remodeling of tissue-engineered devices in vivo in individual patients through molecular imaging technology [123]. Comprising in vivo characterization and measurement of biologic processes at the cellular and molecular level using noninvasive, high-resolution, in vivo imaging technology, molecular imaging detects molecular processes and has the capability to assess gene expression in vivo [124]. Ultimately, molecular imaging techniques might be used in vivo to non-invasively monitor the quality of tissue healing and remodeling of implanted tissue-engineered heart valves.

Manufacturers do not currently possess the skills and infrastructure necessary for commercializing engineered products. New devices have generally been evaluated through standard in vitro testing methods, such as hydrodynamic and accelerated fatigue testing. However, these methods when used for evaluation of valve design would provide only initial information, as valve performance will change and should improve as a degradable scaffold is remodeled. Until now, testing methods were never required to provide quality control of a biologically active product. Quality control should quantitatively assess the concentration of cell attachment, extent of cell coverage, the morphology and phenotype of seeded cells, and their bioactivity.

Nevertheless, conventional testing methods are not adequate to address the biological response elicited by valve or validate bioactive components in vivo. Development of quantitative assays for quality control is a key challenge. These methods must be able to quantitatively determine the concentration of the attached factor of interest or its bioactivity on cell population or scaffold. Maintenance of cell lines and regulation of the seeding process are major issues. If bioreactors are a necessary step in the process, they will need to be scaled-up for multiple valve use while assuring consistency of exposure to different parameters. Potential modes of failure for engineered valves must also be determined. Regulatory agencies must develop standards and guidelines on testing that are required to demonstrate the safety, efficacy, and performance of these products. Another major challenge is that animal models yielding results that can be translated to predictions for humans will be needed.

11 Conclusions

Tissue-engineering approaches to repair and regenerate cardiovascular tissues in general, and heart valves in particular, are imaginative and hold great potential, but efforts have been largely empirical to date. Key scientific inputs are necessary. Future directions include new ways to synthesize and fabricate biomaterials, new ways to control the human body's response to materials, and rational design of materials, which interact with specific molecular or cellular components in the body. These approaches and opportunities created by exciting advances in cell biology raise exciting possibilities for in situ regeneration and repair of cardiovascular tissues (including, potentially, an entire heart) [125].

References

1. Vongpatanasin W, Hillis D, Lange RA (1996) N Engl J Med 335:407–416
2. Bloomfield P, Wheatley DJ, Prescott RJ, Miller HC (1991) N Engl J Med 324:573–579
3. Hammermeister KE, Sethi GK, Henderson WG, Grover FL, Oprian C, Rahimtoola S (2000) J Am Coll Cardiol 36:1152–1158
4. Schoen FJ, Levy RJ (1999) J Biomed Mater Res 47:439–465
5. Schoen FJ (2004) The heart. In: Robbins & Cotran Pathologic Basis of Disease, 7th Ed., Kumar V, Fausto N, Abbas a (eds.), Philadelphia, WB Saunders, pp 555–618
6. Mjaatvedt CH, Yamamura H, Wessels A, Ramsdell A, Turner D, Markwald RR (1999) Mechanisms of segmentation, septation, and remodeling of the tubular heart: Endocardial cushion fate and cardiac looping. In: Harvey RP, Rosenthal N (eds) Heart development. Academic, San Diego, pp 159–177
7. Eisenberg LM, Markwald RR (1995) Circ Res 77:1–7
8. Jiang X, Rowitch DH, Soriano P, McMahon AP, Sucov HM (2000) Development 127: 1607–1616
9. Markwald RR, Fitzharris TP, Manasek FJ (1977) Am J Anat 148:85–129
10. Beresford WA (1999) Transdifferentiation and the vascular wall. In: Zilla P, Greisler HP (eds) Tissue engineering of vascular prosthetic grafts. Landes, Georgetown, TX, p 403
11. Johnson EN, Lee YM, Sander TL, Rabkin E, Schoen FJ, Kaushal S, Bischoff J (2003) J Biol Chem 278:1686–1692
12. DeRuiter MC, Poelmann RE, VanMunsteren JC, Mironov V, Markwald RR, Gittenberger-de Groot AC (1997) Circ Res 80:444–451
13. Thiery JP (2003) Curr Opin Cell Biol 15:740–746
14. Savagner P (2001) BioEssays 23:912–923
15. Boyer AS, Ayerinskas II, Vincent EB, McKinney LA, Weeks DL, Runyan RB (1999) Dev Biol 208:530–545
16. Brown CB, Boyer AS, Runyan RB, Barnett JV (1999) Science 283:2080–2082
17. Camenisch TD, Schroeder JA, Bradley J, Klewer SE, McDonald JA (2002) Nat Med 8:850–855
18. Lee TC, Zhao YD, Courtman DW, Stewart DJ (2000) Circulation 101:2345–48
19. Gitler AD, Epstein JA (2003) Cell Cycle 2:96–98
20. de la Pompa JL, Timmerman LA, Takimoto H et al. (1998) Nature 392:182–185

21. Ranger AM, Grusby MJ, Hodge MR et al. (1998) Nature 392:186–189
22. Camenisch TD, Schroeder JA, Bradley J, Klewer SE, McDonald JA (2002) Nat Med 8:850–855
23. Schoen FJ (1997) J Heart Valve Dis 6:1–6
24. Kunzelman KS, Cochran RP, Murphree SS, Ring WS, Verrier ED, Eberhart RC (1993) J Heart Valve Dis 2:236–244
25. Scott M, Vesely I (1995) Ann Thorac Surg 60:5391–5394
26. Shin D et al. (2001) Dev Biol 230:139
27. Rabkin-Aikawa E, Aikawa M, Farber M, Krats J, Garcia-Cardena G, Kouchoukos NT, Mitchell MB, Jonas RA, Schoen FJ (2004) J Thorac Cardiovasc Surg 128:552-561
28. Butcher JT, Penrod AM, Garcia AJ, Nerem RM (2004) Arterioscler Thromb Vasc Biol 24:1429–1434
29. Paranya G, Vineberg S, Dvorin E, Kaushal S, Roth SJ, Rabkin E, Schoen FJ, Bischoff J (2001) Am J Pathol 159:1335–43
30. Mitchell RN, Jonas RA, Schoen FJ (1998) J Thorac Cardiovasc Surg 115:118–127
31. Chen RH, Kadner A, Mitchell RN, Adams DH (2000) J Thorac Cardiovasc Surg 119: 1216–1220
32. Muller AM, Cronen C, Kupferwasser LI, Oelert H, Muller KM, Kirkpatrick CJ (2000) J Pathol 191:54–60
33. Ghaisas NK, Foley JB, O'Briain S, Crean P, Kelleher D, Walsh M (2000) J Am Coll Cardiol 36:2257–2262
34. Dvorin EL, Jacobson J, Roth SJ, Bischoff J (2003) J Heart Valve Dis 12:617–624
35. Della Rocca F, Sartore S, Guidolin D, Bertiplaglia B, Gerosa G, Casarotto D, Pauletto P (2000) Ann Thorac Surg 70:1594–1600
36. Taylor PM, Allen SP, Yacoub MH (2000) J Heart Valve Dis 9:150–158
37. Taylor PM, Batten P, Brand NJ, Thomas PS, Yacoub MH (2003) Int J Biochem Cell Biol 35:113–118
38. Roy A, Brand NJ, Yacoub MH (2000) J Heart Valve Dis 9:459–465
39. Chester AH, Misfeld M, Yacoub MH (2000) J Heart Valve Dis 9:250–255
40. Hafizi S, Taylor PM, Chester AH, Allen SP, Yacoub MH (2000) J Heart Valve Dis 9:454–458
41. Schurch W, Seemayer TA, Gabbiani G (1998) Am J Surg Pathol 22:141–7
42. Schmitt-Graff A, Desmouliere A, Gabbiani G (1994) Virchows Arch 425:3–24
43. Aikawa M, Sivam PN, Kuro-o M et al. (1993) Circ Res 73:1000–1012
44. Lester WM, Damji AA, Tanaka M, Gedeon I (1992) J Mol Cell Cardiol 24:43–53
45. Lester WM, Damji AA, Gedeon I, Tanaka M (1993) In Vitro Cell Dev Biol 29A:41–50
46. Meredith JEJ, Schwartz MA (1997) Trends Cell Biol 7:146–151
47. Woodard AS, Garcia-Cardena G, Cheresh DA, Leong M, Madri JA, Sessa WC, Languino LR (1998) J Cell Sci 111:469–478
48. Durbin AD, Gotlieb AI (2002) Cardiovasc Pathol 11:69–77
49. Schoen FJ (1999) J Heart Valve Dis 8:350–358
50. Rabkin E, Hoerstrup SP, Aikawa M, Mayer JE Jr, Schoen FJ (2002) J Heart Valve Dis 11:308–314
51. Tamura K, Jones M, Yamada I, Ferrans VJ (2000) J Heart Valve Dis 9:53–63
52. Rabkin-Aikawa E, Farber M, Aikawa M, Schoen FJ (2004) J Heart Valve Dis 13:841–847
53. Visse R, Nagase H (2003) Circ Res 92:827–839
54. Krane SM, Byrne MH, Lemaitre V et al. (1996) J Biol Chem 271:28509–28515
55. Sukhova GK, Shi G-P, Simon DI et al. (1998) J Clin Invest 102:576–583
56. Decker RS, Dingle JT (1982) Science 215:987–989
57. Dreger SA, Taylor PM, Allen SP, Yacoub MH (2002) J Heart Valve Dis 11:875–880
58. Rabkin E, Aikawa M, Stone JR, Fukumoto Y, Libby P, Schoen FJ (2001) Circulation 104:2525–2532

59. Lester W, Rosenthal A, Granton B, Gotlieb AI (1988) Porcine mitral valve interstitial cells in culture. Lab Invest 59:710–719
60. Butany J, Leask R (2001) J Long Term Effects Med Implants 11:115–135
61. Grunkemeier GL, Jamieson WR, Miller DC et al. (1994) J Thorac Cardiovasc Surg 108:709–718
62. Schoen FJ, Hobson CE (1985) Hum Pathol 16:549–559
63. Lester WM, Gotlieb AI (1988) In vitro repair of the wounded porcine mitral valve. Cir Res 62:833–845
64. Valente M, Bortolotti U, Thiene G (1985) Am J Pathol 119:12–21
65. Schoen FJ (1989) Interventional and surgical cardiovascular pathology: clinical correlations and basic principles. Saunders, Philadelphia
66. Schoen FJ, Tsao JW, Levy RJ (1986) Am J Pathol 123:134–145
67. Schoen FJ, Levy RJ, Nelson AC et al. (1985) Lab Invest 52:523–532
68. Sacks MS, Schoen FJ (2002) J Biomed Mater Res 62:359–371
69. Simionescu D, Simionescu A, Deac R (1993) Biomed Mater Res 27:821–829
70. National Institutes of Health (1999) Working group on tissue genesis and organogenesis for heart, lung and blood applications. August 13, 1999. http://www.nhlbi.nih.gov/meetings/workshops/tissueg1.html
71. Nerem RM, Seliktar D (2001) Ann Rev Biomed Eng 3:225–243
72. Edelman ER (1999) Circ Res 85:1115–1117
73. Nugent HM, Edelman ER (2003) Circ Res 92:1068–1078
74. Rabkin E, Schoen FJ (2002) Cardiovasc Pathol 11:305–317
75. Meinhart JG, Deutsch M, Fischlein T, Howanietz N, Froschl A, Zilla P (2001) Ann Thorac Surg 71:S327-331
76. Kaushal S, Amiel GE, Guleserian KJ et al. (2001) Nat Med 7:1035–1040
77. El Oakley RM, Ooi OC, Bongso A, Yacoub MH (2001) Ann Thorac Sur 71:1724–1733
78. Menasche P (2001) Semin Thorac Cardiovasc Surg 14:157–166
79. Steinhoff G, Stock U, Karim N, Mertsching H, Timke A et al. (2000) 102:III50–III55
80. O'Brien MF, Goldstein S, Walsh S, Black KS, ElkinsR, Clarke D (1999) Semin Thorac Cardiovasc Surg 11:194–200
81. L'Heureux N, Stoclet JC, Auger FA, Lagaud GJ, German L, Andriantsitohaina R (2001) FASEB J 15:515–524
82. Tien J, Chen CS (2002) IEEE Eng Med Biol 2:95–98
83. Bhadriraju K, Chen CS (2002) Drug Discovery Today 7:612–620
84. Platt JL, Nagayasu T (1999) Clin Exp Pharm Physiol 26:1026–1032
85. Ballas CB, Zielske SP, Gerson SL (2002) J Cell Biochem 38:20–28
86. Asahara T, Murohara T, Sullivan A et al. (1997) Science 275:964–967
87. Prockop DJ (1997SScience 276:71-74
88. Levenberg S, Golub JS, Amit M, Itskovitz-Eldor J, Langer R (2002) Proc Natl Acad Sci 99:4391-4396
89. Robertson JA (2001) Nat Rev Genet 2:74–78
90. De WG, Berghmans RL, Boer GJ, Andersen S, Brambati B, Carvalho A et al. (2002) Med Health Care Philos 5:79–90
91. Reiss MJ (2000) J Appl Philos 17:253–262
92. Masuda H, Kalka C, Asahara T (2000) Hum Cell 13:153–160
93. Shimizu K, Sugiyama S, Aikawa M, Fukumoto Y, Rabkin E, Libby P, Mitchell RN (2001) Nat Med 7:378–341
94. Caplice NM, Bunch TJ, Stalboerger PG, Wang S, Simper Dmiller DV et al. (2003) Proc Natl Acad Sci 100:4754–4759
95. Rafii S, Lyden D (2003) Nat Med 9:702–712

96. Chaikof EL, Matthew H, Kohn J, Mikos AG, Prestwich GD, Yip CM (2002) Ann NY Acad Sci 961:96–105
97. Mikos AG, Lu L, Temenoff JS, Tessmar JK (in press) Synthetic bioresorbable polymer scaffolds. In: Ratner BD, Hoffman AS, Schoen FJ, Lemons JE (eds) Biomaterials science: an introduction to materials in medicine, 2nd edn. Academic, San Diego
98. Hodde J (2002) Tissue Eng 8:295–308
99. Schmidt CE, Baier JM (2000) Biomaterials 21:2215–2231
100. Kim B, Mooney DJ (1998) Trends Biotechnol 16:224–230
101. Pins GD, Toner M, Morgan JR (2000) FASEB J 14:593–602
102. Williams SF, Martin DP, Horowitz DM, Peoples OP (1999) Int J Biol Macromol 25: 111–121
103. Lendlein A, Langer R (2002) Science 296:1673–1676
104. Grodzinsky AJ, Levenston ME, Jin M, Frank EH (2000) Ann Rev Biomed Eng 2: 691–713
105. Hoerstrup SP, Sodian R, Daebritz S et al. (2000) Circulation 102:III44–49
106. Taylor P, Allen SP, Dreger SA, Yacoub MH (2002) J Heart Valve Dis 11:298–307
107. Curtil A, Pegg DE, Wilson A (1997) J Heart Valve Dis 6:296–306
108. Bader A, Schilling T, Teebken OE, Brandes G, Herden T, Steinhoff G et al. (1998) Eur J Cardiothorac Surg 14:279–284
109. Steinhoff G, Stock U, Karim N, Mertsching H, Timke A et al. (2000) Circulation 102: III50–III55
110. Bengtsson LA, Phillips R, Haegerstrand AN (1995) Ann Thorac Surg 60:S365–368
111. Elkins RC, Goldstein S, Hewitt CW, Walsh SP, Dawson PE et al. (2001) Semin Thorac Cardiovasc Surg 13:87–92
112. O'Brien MF, Goldstein S, Walsh S, Black KS, Elkins R, Clarke D (1999) Semin Thorac Cardiovasc Surg 11:194–200
113. Elkins RC, Dawson PE, Goldstein S, Walsh SP, Black KS (2001) Ann Thorac Surg 71: S428–432
114. Sievers HH, Stierle U, Schmidtke C, Bechtel M (2003) Z Kardiol 92:53–59
115. Simon P, Kasimirr MT, Seebacher G, Weidel G, Ullrich R, Salzer-Muhar U, Rieder E, Wolner E (2003) Eur J Cardiothorac Surg 23:1002–1006
116. Matheny RG, Hutchison ML, Dryden PE et al. (2000) J Heart Valve Dis 9:769–775
117. Rahimtoola SH (2003) J Am Coll Cardiol 42:1720–1721
118. Weinshilboum R (2003) N Engl J Med 348:529–537
119. Ye S (2000) Matrix Biol 19:623–239
120. Jones GT, Phillips VL, Harris EL, Rossaak JI, van Rij AM (2003) J Vasc Surg 38:1363–1367
121. Gencbay M, Turan F, Degertekin M, Eksi N, Mutlu B, Unalp A (1998) J Heart Valve Dis 7:601–609
122. Schwartz RS, Edelman ER (2002) Circulation 106:1867–1873
123. Sameni M, Dosescu J, Moin K, Sloane BF (2003) Mol Imaging 2:159–175
124. Weissleder R, Mahmood U (2001) Mol Imaging Radiol 219:316–333
125. Mann BK, West JL (2001) Anat Rec 263:367–371
126. Markwald G et al. (1975) Dev Biol 42:160–180
127. Schoen FJ, Edwards WD (2001) Valvular heart disease: general principles and stenosis. In: Silver MD, Gotlieb AI, Schoen FJ (eds) Cardiovascular pathology, 3rd edn. Saunders, Philadelphia, pp 402–442
128. Langer R, Vacanti JP (1993) Science 56:423–427
129. Massia SP (1999) Cell-extracellular matrix interactions relevant to vascular tissue engineering. In: Zilla P, Greisler HP (eds) Tissue engineering of vascular prosthetic grafts. Landes, Georgetown, TX, 1999, p 583–597
130. Harken DE et al. (1962) Am J Cardiol 9:292–299

131. Schroeder JA, Jackson LF, Lee DC, Camenisch TD (2003) Form and function of developing heart valve: coordination by extracellular matrix and growth factor signaling. J Mol Med 81:392–403
132. Lincoln J, Alfieri CM, Yutzey KE (2004) Development of heart valve leaflets and supporting apparatus in chicken and mouse embryos. Develop Dynamics 230:239–250
133. Armstrong EJ, Bischoff J (2004) Heart valve development: endothelial cell signaling and differentiation. Circ Res 95:459–470
134. Norris RA, Kern CB, Wessels A, Moralez EI, Markwald RR, Mjaatvedt CH (2004) Identification and detection of the periostin gene in cardiac development. Anat Rec 281A:1227–1233
135. Davies PF, Passerini AG, Simmons CA (2004) Aortic valve: turning over a new leaf(let) in endothelial phenotype heterogeneity. Arterioscler Thromb Vasc Biol 24:1331–1333
136. Walker GA, Masters KS, Shah DN, Anseth KS, Leinwand LA (2004) Valvular myofibroblast activation by transforming growth factor beta: implications for pathological extracellular matrix remodeling in heart valve disease. Cic Res 95:253–260

Received: May 2004

Adv Biochem Engin/Biotechnol (2005) 91: 181–210
DOI 10.1007/b100004

Regeneration of Urologic Tissues and Organs

Anthony Atala (✉)

The William Boyce Professor and Chair, Department of Urology, Director,
Wake Forest Institute for Regenerative Medicine, Wake Forest University School
of Medicine, Winston Salem, NC 27157, USA
aatala@wfubmc.edu

Abstract Patients suffering from a variety of urologic diseases may be treated with transplanted tissues and organs. However, there is a shortage of donor tissues and organs, which is worsening yearly owing to the ageing population. Scientists in the field of regenerative medicine and tissue engineering are applying the principles of cell transplantation, material science, and bioengineering to construct biological substitutes that will restore and maintain normal function in diseased and injured urologic tissues. This chapter reviews recent advances that have occurred in the regeneration of urologic organs and describes how these applications may offer novel therapies for patients with urologic disease.

Keywords Kidney · Bladder · Urethra · Genitalia · Cloning

Abbreviations

AA	Amino acid
AQP1	Aquaporin 1
AQP2	Aquaporin 2
CTL	Cytotoxic T lymphocytes
DTH	Delayed-type hypersensitivity
ECM	Extracellular matrix
miHA	Minor histocompatibility antigens
MLC	Mixed lymphocyte culture
PGA	Polyglycolic acid
PLA	Polylactic acid
PLGA	Poly(lactic-*co*-glycolic acid)
SCNT	Somatic cell nuclear transfer

1 Introduction

After flying the first solo transatlantic flight, Charles Lindbergh spent the rest of his life working with Alexis Carrell, a Nobel Laureate in medicine, at the Rockefeller Institute in New York culturing organs in the hope of using tissues for future transplantation. Together, Carrell and Lindbergh published a book in 1938, titled *The Culture of Organs* [1]. Organ cultures of urologic tissues were described in their seminal work. Over 60 years later, the culture and expansion of cells for the engineering of various urologic tissues is now routine. Over the last decade, the ability to engineer urologic tissues has broadened the theoretical options regarding the future of genitourinary reconstruction.

The genitourinary system, consisting of the kidney, ureter, bladder, prostate, urethra, male and female genitalia and reproductive structures, is exposed to a variety of possible injuries and anomalies from the time the fetus develops. Aside from congenital abnormalities, individuals may also suffer from other disorders such as cancer, trauma, infection, inflammation, iatrogenic injuries or other conditions that may lead to genitourinary organ damage or loss, requiring eventual reconstruction. The type of tissue chosen for replacement depends on which organ requires reconstruction. Bladder and ureteral reconstruction may be performed with gastrointestinal tissues. Urethral reconstruction is performed with skin, mucosal grafts from the bladder, rectum or oral cavity. Vaginas can be reconstructed with skin, small bowel, sigmoid colon, and rectum. However, a shortage of donor tissue may limit these types of reconstructions and there is a degree of morbidity associated with the harvest procedure. In addition, these approaches rarely replace the entire function of the original organ. The tissues used for reconstruction may lead to complications owing to their inherently different functional parameters. In most cases, the replacement of lost or deficient tissues with functionally equivalent tissues would improve the outcome for these patients. This goal may be attainable with the use of tissue engineering techniques.

Tissue engineering, a component of regenerative medicine, follows the principles of cell transplantation, materials science and engineering towards the development of biological substitutes which would restore and maintain normal function. Tissue engineering may involve matrices alone, wherein the body's natural ability to regenerate is used to orient or direct new tissue growth, or the use of matrices with cells. When cells are used for tissue engineering, donor tissue is dissociated into individual cells, which are either implanted directly into the host or expanded in culture, attached to a support matrix, and re-implanted after expansion. The implanted tissue can be heterologous, allogeneic, or autologous. Ideally, this approach might allow lost tissue-function to be restored or replaced in toto and with limited complications [2]. The use of autologous cells would avoid rejection, wherein a biopsy of tissue is obtained from the host, the cells are dissociated and expanded in vitro, reattached to a matrix, and implanted into the same host [2–21].

2
Cell Growth

One of the initial limitations of applying cell-based tissue engineering techniques to urologic organs has been the previously encountered inherent difficulty of growing genitourinary associated cells in large quantities. In the past, it was believed that urothelial cells had a natural senescence that was hard to overcome. Normal urothelial cells could be grown in the laboratory setting, but with limited expansion. Several protocols have been developed over the last two decades that have improved urothelial growth and expansion [10, 22–24]. A system of urothelial cell harvest was developed that does not use any enzymes or serum and has a large expansion potential. Using these methods of cell culture, it is possible to expand a urothelial strain from a single specimen which initially covers a surface area of 1 cm^2 to one covering a surface area of 4,202 m^2 (the equivalent area of one football field) within 8 weeks [10]. These studies indicated that it should be possible to collect autologous urothelial cells from human patients, expand them in culture, and return them to the human donor in sufficient quantities for reconstructive purposes. Bladder, ureter and renal pelvis cells can be equally harvested, cultured and expanded in a similar fashion. Normal human bladder epithelial and muscle cells can be efficiently harvested from surgical material, extensively expanded in culture, and their differentiation characteristics, growth requirements and other biological properties studied [10, 16, 23–32].

3 Biomaterials

Biomaterials provide a cell-adhesion substrate and can be used to achieve cell delivery with high loading and efficiency to specific sites in the body. The configuration of the biomaterials can guide the structure of an engineered tissue. The biomaterials provide mechanical support against in vivo forces, thus maintaining a predefined structure during the process of tissue development. The biomaterials can be loaded with bioactive signals, such as cell-adhesion peptides and growth factors, which can regulate cellular function. The design and selection of the biomaterial is critical in the development of engineered genitourinary tissues. The biomaterial must be capable of controlling the structure and function of the engineered tissue in a predesigned manner by interacting with transplanted cells and/or the host cells. Generally, the ideal biomaterial should be biocompatible, promote cellular interaction and tissue development, and possess proper mechanical and physical properties.

The selected biomaterial should be biodegradable and bioresorbable to support the reconstruction of a completely normal tissue without inflammation. The degradation products should not provoke inflammation or toxicity and must be removed from the body via metabolic pathways. The degradation rate and the concentration of degradation products in the tissues surrounding the implant must be at a nearly physiological level [33]. The mechanical support of the biomaterials should be maintained until the engineered tissue has sufficient mechanical integrity to support itself [4]. This can be potentially achieved by an appropriate choice of mechanical and degradative properties of the biomaterials [34].

Generally, three classes of biomaterials have been utilized for engineering genitourinary tissues: naturally-derived materials (e.g., collagen and alginate), acellular tissue matrices (e.g., bladder submucosa and small intestinal submucosa), and synthetic polymers [e.g., polyglycolic acid (PGA), polylactic acid (PLA), and poly(lactic-*co*-glycolic acid) (PLGA)]. These classes of biomaterials have been tested in respect to their biocompatibility with primary human urothelial and bladder muscle cells [35, 36]. Naturally derived materials and acellular tissue matrices have the potential advantage of biological recognition. Synthetic polymers can be produced reproducibly on a large scale with controlled properties of strength, degradation rate, and microstructure.

Collagen is the most abundant and ubiquitous structural protein in the body, and may be readily purified from both animal and human tissues with an enzyme treatment and salt/acid extraction [37]. Collagen implants degrade through a sequential attack by collagenases. The in vivo resorption rate can be regulated by controlling the density of the implant and the extent of intermolecular crosslinking. The lower the density, the greater the interstitial space, and generally, the larger the pores for cell infiltration, leading to a higher rate of implant degradation. Collagen contains cell-adhesion domain sequences (e.g., RGD), which exhibit specific cellular interactions. This may assist in retaining the phenotypes and activities of many types of cells, including fibroblasts [38] and chondrocytes [39].

Alginate, a polysaccharide isolated from sea weed, has been used as an injectable cell delivery vehicle [40] and a cell immobilization matrix [41] owing to its gelling properties in the presence of divalent ions such as calcium. Alginates are relatively biocompatible and approved by the FDA for human use as wound dressing material. Alginate is a family of copolymers of D-mannuronate and L-guluronate. The physical and mechanical properties of alginate gel are strongly correlated with the proportion and length of polyguluronate block in the alginate chains [40].

Acellular tissue matrices are collagen-rich matrices prepared by removing cellular components from tissues. The matrices are often prepared by mechanical and chemical manipulation of a segment of tissue [15, 42–44]. The matrices slowly degrade upon implantation, and are replaced and remodeled by ECM proteins synthesized and secreted by transplanted or ingrowing cells.

Polyesters of naturally occurring α-hydroxy acids, including PGA, PLA, and PLGA, are widely used in tissue engineering. These polymers have gained FDA approval for human use in a variety of applications, including sutures [45]. The ester bonds in these polymers are hydrolytically labile, and these polymers degrade by nonenzymatic hydrolysis. The degradation products of PGA, PLA, and PLGA are nontoxic, natural metabolites and are eventually eliminated from the body in the form of carbon dioxide and water [45]. The degradation rate of these polymers can be tailored from several weeks to several years by altering crystallinity, initial molecular weight, and the copolymer ratio of lactic to glycolic acid. Since these polymers are thermoplastics, they can be easily formed into a three-dimensional scaffold with the desired microstructure, gross shape and dimensions by various techniques, including molding, extrusion [46], solvent-casting [47], phase separation techniques, and gas-foaming techniques [48]. Many applications in genitourinary tissue engineering often require a scaffold with high porosity and ratio of surface area to volume. Other biodegradable synthetic polymers, including poly(anhydrides) and poly(ortho-esters), can also be used to fabricate scaffolds for genitourinary tissue engineering with controlled properties [49].

4 Regeneration of Urologic Structures

4.1 Urethra

Various biomaterials without cells have been used experimentally (in animal models) for the regeneration of urethral tissue, including PGA, and acellular collagen-based matrices from small intestine, and bladder [44, 50–54]. Some of these biomaterials, such as acellular collagen matrices derived from bladder submucosa, have also been seeded with autologous cells for urethral reconstruction. Our laboratory has been able to replace tubular urethral segments with cell-seeded collagen matrices.

Acellular collagen matrices derived from bladder submucosa in our laboratory have been used experimentally and clinically. In animal studies, segments of the urethra were resected and replaced with acellular matrix grafts in an onlay fashion. Histological examination showed complete epithelialization and progressive vessel and muscle infiltration. The animals were able to void through the neourethras [44]. These results were confirmed clinically in a series of patients with hypospadias, an anatomic anomaly in which the urethral opening is not properly located, and urethral stricture disease [21, 55]. Cadaveric bladders were microdissected and the submucosal layers were isolated. The submucosa was washed and decellularized. The matrix was used for urethral repair in patients with stricture disease (n=33; 28 adults, 5 children) and hypospadias (n=7 children). The matrices were trimmed to size and the neo-urethras were created by anastomosing the matrix in an onlay fashion to the urethral plate. The size of the neourethras ranged from 2 to 16 cm. Voiding histories, physical examination, retrograde urethrography, uroflowmetry, and cystoscopies were performed serially, pre- and post-operatively, with up to a 7 year follow-up. After a 4–7 year follow-up, 34 of the 40 patients had a successful outcome. Six patients with a urethral stricture had a recurrence, and one patient with hypospadias developed a fistula, an opening along the newly developed urinary channel. The mean maximum urine flow rate significantly increased post-operatively. Cystoscopic studies showed adequate caliber conduits. Histologic examination of the biopsies showed the typical urethral epithelium. The use of an off-the-shelf matrix appears to be beneficial for patients with abnormal urethral conditions, and obviates the need for obtaining autologous grafts, thus decreasing operating time and eliminating donor site morbidity (Fig. 1).

Unfortunately, the use of acellular matrices is not applicable for tubularized urethral repairs (the entire urethral circumference), as opposed to onlay repairs (only a portion of the urethral circumference). The collagen matrices are able to replace urethral segments when used in an onlay fashion. However, if a tubularized repair is needed, the collagen matrices need to be seeded with autologous cells [56, 57]. Autologous bladder epithelial and smooth muscle cells from male rabbits were grown and seeded onto pre-configured tubular matrices. The

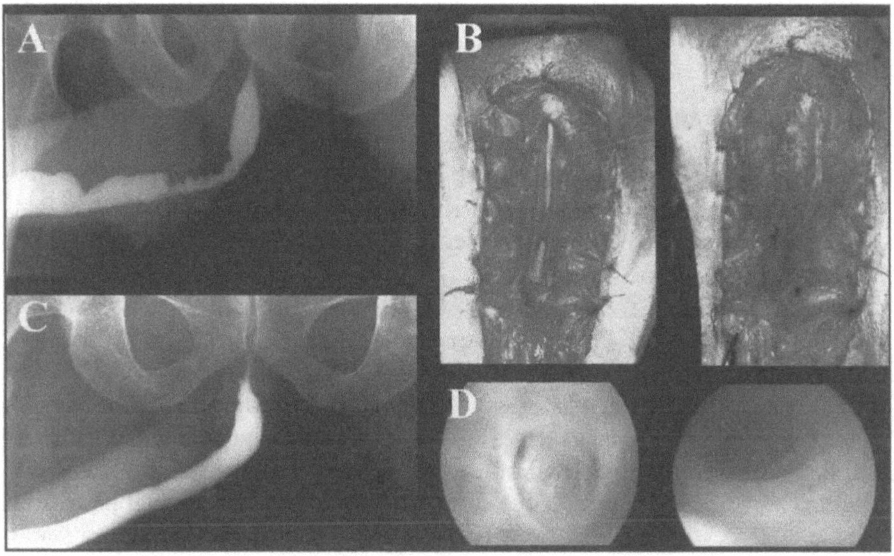

Fig. 1 A–D Representative case of a patient with a bulbar stricture repaired with a collagen matrix. **A**, preoperative urethrogram. **B**, urethral repair. Strictured tissue is excised, preserving urethral plate on left side and matrix is anastomosed to urethral plate in an onlay fashion on right side. **C**, urethrogram 6 months after repair. **D**, cystoscopic view of urethra preoperatively on left side and 4 months after repair on right side.

entire anterior urethra was resected and urethroplasties were performed with tubularized collagen matrices seeded with cells in nine animals, and without cells in six animals. Serial urethrograms showed a wide urethral caliber without strictures in the animals implanted with the cell-seeded matrices, and collapsed urethral segments with strictures within the unseeded scaffolds. Gross examination of the urethral implants seeded with cells showed normal-looking tissue without any evidence of fibrosis. Histologically, a transitional cell layer surrounded by muscle cell fiber bundles with increasing cellular organization over time were observed on the cell seeded constructs. The epithelial and muscle phenotypes were confirmed with pAE1/AE3 and smooth-muscle-specific α-actin antibodies. A transitional cell layer with scant unorganized muscle fiber bundles and large areas of fibrosis were present at the anastomotic sites on the unseeded constructs. Therefore, tubularized collagen matrices seeded with autologous cells can be used successfully for total penile urethra replacement; whereas tubularized collagen matrices without cells lead to poor tissue development and stricture formation. The cell-seeded collagen matrices form new tissue which is histologically similar to native urethra [56, 57]. This technology may be applicable to patients requiring tubularized urethral repair.

5 Bladder

Currently, gastrointestinal segments are commonly used as tissues for bladder replacement or repair. However, gastrointestinal tissues are designed to absorb specific solutes, whereas bladder tissue is designed for the excretion of solutes. Owing to the problems encountered with the use of gastrointestinal segments, numerous investigators have attempted alternative materials and tissues for bladder replacement or repair.

Over the last few decades, several bladder wall substitutes have been attempted with both synthetic and organic materials. The first application of a free tissue graft for bladder replacement was reported by Neuhoff in 1917, when fascia was used to augment bladders in dogs [58]. Since that first report, multiple other free graft materials have been used experimentally and clinically, including bladder allografts, small intestinal submucosa, pericardium, dura, and placenta [15, 42, 59–66]. In multiple studies using different materials as an acellular graft for cystoplasty, the urothelial layer was able to regenerate normally, but the muscle layer, although present, was not fully developed [15, 64, 67, 68]. When using cell-free collagen matrices, scarring and graft contracture may occur over time [69–74]. Synthetic materials, which have been tried previously in experimental and clinical settings, include polyvinyl sponge, tetrafluoroethylene (Teflon), vicryl matrices, and silicone [75–78]. Most of the above attempts have usually failed, owing to mechanical, structural, functional, or biocompatibility problems. Usually, nondegrading synthetic materials used for bladder reconstruction succumb to mechanical failure and urinary stone formation, while degradable materials lead to fibroblast deposition, scarring, graft contracture, and a reduced reservoir volume over time.

Engineering tissue using selective cell transplantation may provide a means to create functional new bladder segments [2]. The success of using cell transplantation strategies for bladder reconstruction depends on the ability to use donor tissue efficiently and to provide the right conditions for long-term survival, differentiation and growth. Urothelial and muscle cells can be expanded in vitro, seeded onto the polymer scaffold, and allowed to attach and form sheets of cells. The cell–polymer scaffold can then be implanted in vivo. A series of in vivo urologic associated cell–polymer experiments were performed. Histologic analysis of human urothelial, bladder muscle, and composite urothelial and bladder muscle–polymer scaffolds, implanted in athymic mice and retrieved at different time points, indicated that viable cells were evident in all three experimental groups [9]. Implanted cells oriented themselves spatially along the polymer surfaces. The cell populations appeared to expand from one layer to several layers of thickness with progressive cell organization with extended implantation times. Cell–polymer composite implants of urothelial and muscle cells, retrieved at extended times (50 days), showed extensive formation of multi-layered sheet-like structures and well-defined muscle layers. Polymers seeded with cells and manipulated into a tubular configuration showed layers of muscle cells lining the multilayered epithelial sheets. Cellular debris appeared reproducibly in the luminal spaces, suggesting that epithelial cells lining the lumina are sloughed into the luminal space. Cell polymers implanted with human bladder muscle cells alone showed almost complete replacement of the polymer with sheets of smooth muscle at 50 days. This experiment demonstrated, for the first time, that composite tissue-engineered structures could be created de novo. Prior to this study, only single-cell type tissue-engineered structures had been created.

5.1 Formation of Bladder Tissue ex situ

In order to determine the effects of implanting engineered tissues in continuity with the urinary tract, an animal model of bladder augmentation was utilized [15]. Partial cystectomies, which involved removing approximately 50% of the native bladders, were performed in ten beagles. In five, the retrieved bladder tissue was microdissected and the mucosal and muscular layers separated. The bladder urothelial and muscle cells were cultured using the techniques described above. Both urothelial and smooth muscle cells were harvested and expanded separately. A collagen-based matrix, derived from allogeneic bladder submucosa, was used for cell delivery. This material was chosen for these experiments owing to its native elasticity. Within 6 weeks, the expanded urothelial cells were collected as a pellet. The cells were seeded on the luminal surface of the allogenic bladder submucosa and incubated in serum-free keratinocyte growth medium for 5 days. Muscle cells were seeded on the opposite side of the bladder submucosa and subsequently placed in DMEM supplemented with 10% fetal calf serum for an additional 5 days. The seeding density on the allogenic bladder submucosa was approximately 1×10^7 cells cm^{-2}.

Preoperative fluoroscopic cystography and urodynamic studies were performed in all animals. Augmentation cystoplasty was performed with the matrix with cells in one group, and with the matrix without cells in the second group. The augmented bladders were covered with omentum, the membrane that encloses the bowel, in order to facilitate angiogenesis to the implant. Cystostomy catheters were used for urinary diversion for 10–14 days. Urodynamic studies and fluoroscopic cystography were performed at 1, 2 and 3 months post-operatively. Augmented bladders were retrieved 2 (n=6) and 3 (n=4) months after surgery and examined grossly, histologically and immunocytochemically.

Bladders augmented with the matrix seeded with cells showed a 99% increase in capacity compared to bladders augmented with the cell-free matrix, which showed only a 30% increase in capacity. Functionally, all animals showed normal bladder compliance, as evidenced by urodynamic studies; however, the remaining native bladder tissue may have accounted for these results. Histologically, the retrieved engineered bladders contained a cellular organization consisting of a urothelial lined lumen surrounded by submucosal tissue and smooth muscle. However, the muscular layer was markedly more prominent in the cell-reconstituted scaffold [15].

Most of the free grafts (without cells) utilized for bladder replacement in the past have been able to show adequate histology in terms of a well-developed urothelial layer; however, they have been associated with an abnormal muscular layer which may be developed to a variable extent [3, 4]. It has been well-established for decades that the bladder is able to regenerate generously over free grafts. Urothelium is associated with a high reparative capacity [79]. Bladder muscle tissue is less likely to regenerate in a normal fashion. Both urothelial and muscle ingrowth are believed to be initiated from the edges of the normal bladder towards the region of the free graft [80, 81]. Usually, however, contracture or resorption of the graft has been evident. The inflammatory response towards the matrix may contribute to the resorption of the free graft.

It was hypothesized that building the three-dimensional structure constructs in vitro, prior to implantation, would facilitate the eventual terminal differentiation of the cells after implantation in vivo, and would minimize the inflammatory response towards the matrix, thus avoiding graft contracture and shrinkage. This study demonstrated that there was a major difference evident between matrices used with autologous cells (tissue-engineered) and matrices used without cells [15]. Matrices implanted with cells for bladder augmentation retained most of their implanted diameter, as opposed to matrices implanted without cells for bladder augmentation, wherein graft contraction and shrinkage occurred. The histomorphology demonstrated a marked paucity of muscle cells and a more aggressive inflammatory reaction in the matrices implanted without cells. Of interest is that the urothelial cell layers appeared normal, even though the underlying matrix was significantly inflamed. It was further hypothesized that having an adequate urothelial layer from the outset would limit the amount of urine contact with the matrix, and would therefore decrease the inflammatory response, and that the muscle cells were also necessary for bioengineering, since native muscle cells are less likely to regenerate over the free grafts. Further studies confirmed this hypothesis [20]. Thus, it appears that the presence of both urothelial and muscle cells on the matrices used for bladder replacement appear to be important for successful tissue bioengineering.

5.2 Bladder Replacement Using Tissue Engineering

The results of initial studies showed that the creation of artificial bladders may be achieved in vivo; however, it could not be determined whether the functional parameters noted were owing to the engineered segment or to the intact native bladder tissue. In order to better address the functional parameters of tissue-engineered bladders, an animal model was designed that required a subtotal cystectomy with subsequent replacement with a tissue-engineered organ [20].

A total of 14 beagle dogs underwent a trigone-sparing cystectomy. The animals were randomly assigned to one of three groups. Group A (n=2) underwent closure of the trigone without a reconstructive procedure. Group B (n=6) underwent reconstruction with a cell-free bladder-shaped biodegradable polymer. Group C (n=6) underwent reconstruction using a bladder-shaped biodegradable polymer that delivered autologous urothelial cells and smooth muscle cells. The cell populations had been separately expanded from a previously harvested autologous bladder biopsy. Preoperative and postoperative urodynamic and radiographic studies were performed serially. Animals were sacrificed at 1, 2, 3, 4, 6 and 11 months postoperatively. Gross, histological and immunocytochemical analyses were performed [20].

The cystectomy-only controls and polymer-only grafts maintained average capacities of 22% and 46% of preoperative values, respectively. An average bladder capacity of 95% of the original pre-cystectomy volume was achieved in the tissue-engineered bladder replacements. These findings were confirmed radiographically. The subtotal cystectomy reservoirs, which were not reconstructed, and polymer-only reconstructed bladders showed a marked decrease in bladder compliance (10% and 42%). The compliance of the tissue-engineered bladders showed almost no difference from preoperative values that were measured when the native bladder was present (106%). Histologically, the polymer-only bladders presented a pattern of normal urothelial cells with a thickened fibrotic submucosa and a thin layer of muscle fibers. The retrieved tissue-engineered bladders showed a normal cellular organization, consisting of a tri-layer of urothelium, submucosa and muscle (Fig. 2). Immunocytochemical analyses for desmin, α-actin, cytokeratin 7, pancytokeratins AE1/AE3 and uroplakin III confirmed the muscle and urothelial phenotype. S-100 staining indicated the presence of neural structures. The results from this study

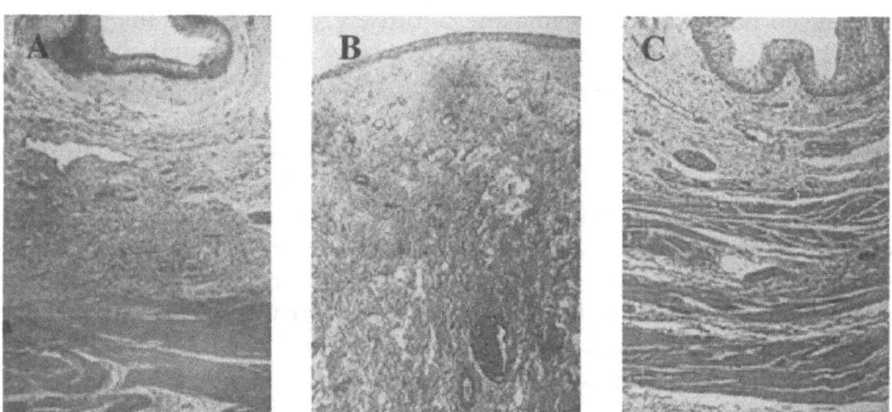

Fig. 2A–C Hematoxylin and Eosin histological results six months after surgery (original magnification: ×250). **A** Normal canine bladder. **B** The bladder dome of the cell-free polymer reconstructed bladder consists of a thickened layer of collagen and fibrotic tissue. **C** The tissue engineered neo-organ shows a histo-morphologically normal appearence. A trilayered architecture consisting of urothelium, submucosa and smooth muscle is evident.

showed that it is possible to tissue-engineer bladders, which are anatomically and functionally normal. Twenty clinical trials for the application of this technology are currently being arranged.

6 Genital Tissues

Reconstructive surgery is required for a wide variety of pathologic penile conditions, such as penile carcinoma, trauma, severe erectile dysfunction and congenital conditions such as ambiguous genitalia, hypospadias and epispadias. One of the major limitations of phallic reconstructive surgery is the availability of sufficient autologous tissue. Phallic reconstruction using autologous tissue, derived from the patient's own cells, may be preferable in selected cases.

7 Corporal Tissues

7.1 Reconstruction of Corporal Smooth Muscle

One of the major components of the phallus is corporal smooth muscle. The creation of autologous functional and structural corporal tissue de novo would be beneficial.

Initial experiments were performed in order to determine the feasibility of creating corporal tissue in vivo using cultured human corporal smooth muscle cells seeded onto PGA scaffolds [82]. Corpus cavernosal smooth muscle cells were isolated from normal young adult patients after informed consent during routine penile surgery. Muscle cells were maintained in culture, seeded onto biodegradable polymer scaffolds, and implanted subcutaneously in athymic mice. Implants were retrieved at 7, 14 and 24 days after surgery for analyses. Corporal smooth muscle tissue was identified, grossly and histologically. Intact smooth muscle cell multilayers were observed growing along the surface of the polymers throughout all time-points. Early vascular ingrowth at the periphery of the implants was evident by 7 days. By 24 days, there was evidence of polymer degradation. Smooth muscle phenotype was confirmed immunocytochemically and by Western blot analyses with antibodies to α-smooth muscle actin.

In order to engineer functional corpus cavernosum, both smooth muscle and sinusoidal endothelial cells are essential. However, penile sinusoidal endothelial cells had not been extensively cultured in the past, and had not been fully characterized. A method of isolation and expansion of sinusoidal endothelial cells from corpora cavernosa was devised, and cell function and gene expression were characterized.

When grown on collagen, corporal cavernosal endothelial cells formed capillary structures which created a complex three-dimensional capillary network. The possibility was investigated of developing human corporal tissue in vivo by combining smooth muscle and endothelial cells [19]. Human corpus cavernosal smooth muscle cells and ECV 304 human endothelial cells were seeded on biodegradable polymer scaffolds and implanted in the subcutaneous space of athymic mice. At retrieval, all polymer scaffolds seeded with cells had formed distinct tissue structures and maintained their pre-implantation size. The control scaffolds without cells had decreased in size with increasing time. Histologically, all of the retrieved polymers seeded with corporal smooth muscle and endothelial cells showed the survival of the implanted cells. The presence of penetrating native vasculature was observed 5 days after implantation. The formation of multilayered strips of smooth muscle adjacent to endothelium was evident by 7 days after implantation. Increased smooth muscle organization and accumulation of endothelium lining the luminal structures were evident 14 days after implantation. A well-organized construct, consisting of muscle and endothelial cells, was noted at 28 and 42 days after implantation. A marked degradation of the polymer fibers was observed by 28 days. There was no evidence of tissue formation in the controls (polymers without cells). The results of these studies suggested that the creation of well-vascularized autologous corporal-like tissue, consisting of smooth muscle and endothelial cells, may be possible.

The aim of phallic reconstruction is to achieve structurally and functionally normal genitalia. It had been shown that human cavernosal smooth muscle and endothelial cells seeded on polymers would form tissue composed of corporal cells when implanted in vivo. However, corporal tissue structurally identical to the native corpus cavernosum was not achieved, owing to the type of polymers used. Therefore, a naturally derived acellular corporal tissue matrix that possesses the same architecture as native corpora was developed. The feasibility of developing corporal tissue, consisting of human cavernosal smooth muscle and endothelial cells in vivo, using an acellular corporal tissue matrix as a cell delivery vehicle was explored [83]. Acellular collagen matrices were derived from processed donor rabbit corpora using cell lysis techniques. Human corpus cavernosal muscle and endothelial cells were derived from donor penile tissue, the cells were expanded in vitro, seeded on the acellular matrices, and implanted subcutaneously in athymic mice. Western blot analysis detected α-actin, myosin and tropomyosin proteins from human corporal smooth muscle cells. Expression of muscarinic acetylcholine receptor (mAChR) subtype m4 mRNA was demonstrated by RT-PCR from corporal muscle cells prior to and 8 weeks after seeding. The implanted matrices showed neovascularity into the sinusoidal spaces by 1 week after implantation. Increasing organization of smooth muscle and endothelial cells lining the sinusoidal walls was observed at 2 weeks and continued with time. The matrices were covered with the appropriate cell architecture 4 weeks after implantation. The matrices showed a stable collagen concentration over 8 weeks, as determined by hydroxy-proline quantification.

Immunocyto-chemical studies using α-actin and Factor VIII antibodies confirmed the presence of corporal smooth muscle and endothelial cells, both in vitro and in vivo, at all time points. There was no evidence of cellular organization in the control matrices.

In another study, we attempted to replace entire crossectional segments of both corpora cavernosa in vivo by interposing engineered tissue in rabbits and investigated their structural and functional integrity [84]. Autologous cavernosal smooth muscle and endothelial cells were harvested, expanded and seeded on acellular collagen matrices. The entire crossection of the protruding rabbit phallus (approximately 0.7 cm long; one third of penile shaft) was excised, leaving the urethra intact. Matrices with and without cells were interposed into the excised corporal space. Additional rabbits, without surgical intervention, served as controls. The experimental corporal bodies demonstrated adequate structural and functional integrity by cavernosography and cavernosometry. Mating activity in the animals with the engineered corpora normalized by 3 months. The presence of sperm was confirmed during mating, and was present in all the rabbits with the engineered corpora. Grossly, the corporal implants with cells showed continuous integration of the graft into native tissue. Histologically, sinusoidal spaces and walls, lined with endothelium and smooth muscle, were observed in the engineered grafts. Grafts without cells contained fibrotic tissue and calcifications with sparse corporal elements. Each cell type was identified immunohistochemically and by Western blot analyses. These studies demonstrate that it is possible to engineer autologous functional penile tissue. Our laboratory is currently working on increasing the size of the engineered constructs.

7.2 Penile Prostheses

Although silicone is an accepted biomaterial for penile prostheses, biocompatibility is a concern [85, 86]. The use of a natural prosthesis composed of autologous cells may be advantageous. A feasibility study for creating natural penile prostheses made of cartilage was performed initially [13].

Non-dividing cartilage cells called chondrocytes, harvested from the articular surface of calf shoulders, were isolated, grown and expanded in culture. The cells were seeded onto pre-formed cylindrical polyglycolic acid polymer rods (1 cm in diameter and 3 cm in length). These cell–polymer scaffolds were implanted in the subcutaneous space of 20 athymic mice. Each animal had two implantation sites consisting of a polymer scaffold seeded with chondrocytes and a control (polymer alone). The rods were retrieved at 1, 2, 4 and 6 months post implantation. Biomechanical properties, including compression, tension and bending, were measured on the retrieved structures. Histological analyses were performed to confirm the cellular composition. At retrieval, all of the polymer scaffolds seeded with cells formed milky-white rod shaped solid cartilaginous structures, maintaining their pre-implantation size and shape.

The control scaffolds without cells failed to form cartilage. There was no evidence of erosion, inflammation or infection in any of the implanted cartilage rods.

The compression, tension and bending studies showed that the cartilage structures were readily elastic and could withstand high degrees of pressure. Biomechanical analyses showed that the engineered cartilage rods possessed the mechanical properties required to maintain penile rigidity. The compression studies showed that the cartilage rods were able to withstand high degrees of pressure. A ramp compression speed of 200 $\mu m\ s^{-1}$, applied to each cartilage rod up to 2,000 μm in distance, resulted in 3.8 kg of resistance. The tension relaxation studies demonstrated that the retrieved cartilage rods were able to withstand stress and were able to return to their initial state while maintaining their biomechanical properties. A ramp tension speed of 200 $\mu m\ s^{-1}$ applied to each cartilage rod created a tensile strength of 2.2 kg, which physically lengthened the rods an average of 0.48 cm. Relaxation of tension at the same speed resulted in retraction of the cartilage rods to their initial state. The bending studies performed at two different speeds showed that the engineered cartilage rods were durable, malleable, and were able to retain their mechanical properties. Cyclic compression, performed at rates of 500 $\mu m\ s^{-1}$ and 20,000 $\mu m\ s^{-1}$, demonstrated that the cartilage rods could withstand up to 3.5 kg of pressure at a predetermined distance of 5,000 μm. The relaxation phase of the cyclic compression studies showed that the engineered rods were able to maintain their tensile strength. None of the rods were ruptured during the biomechanical stress relaxation studies. Full details of these testing procedures are available in the original published article by Yoo et al. [13].

Histological examination with hematoxylin and eosin showed the presence of mature and well-formed cartilage in all the chondrocyte–polymer implants. The polymer fibers were progressively replaced by cartilage with time progression. Undegraded polymer fibers were observed at 1 and 2 months after implantation. However, remnants of polymer scaffolds were not present in the cartilage rods at 6 months. Aldehyde fuschin-alcian blue and toluidine blue staining demonstrated the presence of highly sulfated mucopolysaccharides, which are differentiated products of chondrocytes. There was no evidence of cartilage formation in the controls.

In a subsequent study using an autologous system, the feasibility of applying the engineered cartilage rods in-situ was investigated [18]. Autologous chondrocytes harvested from rabbit ear were grown and expanded in culture. The cells were seeded onto biodegradable poly-L-lactic-acid-coated polyglycolic acid polymer rods at a concentration of 50×10^6 chondrocytes cm^{-3}. Eighteen chondrocyte-polymer scaffolds were implanted into the corpora cavernosa spaces of ten rabbits. As controls, two corpora, one each in two rabbits, were not implanted. The animals were sacrificed at 1, 2, 3 and 6 months after implantation. Histological analyses were performed with hematoxylin and eosin, aldehyde fuschin-alcian blue, and toluidine blue staining. All animals tolerated the implants for the duration of the study without any complications. Gross examination at retrieval showed the presence of well-formed milky white cartilage structures within the corpora at 1 month. All polymers were fully degraded by 2 months. There was no evidence of erosion or infection in any of the implant sites. Histological analyses with alcian blue and toluidine blue staining demonstrated the presence of mature and well-formed chondrocytes in the retrieved implants. Subsequent studies were performed assessing the functionality of the cartilage penile rods in vivo long term. To date, the animals have done well, and can copulate and impregnate their female partners without problems. Further functional studies need to be completed before applying this technology to the clinical setting.

8 Injectable Therapies

Both urinary incontinence and vesicoureteral reflux are common conditions affecting the genitourinary system, wherein injectable bulking agents can be used for treatment. There are definite advantages in treating urinary incontinence and vesicoureteral reflux endoscopically. The method is simple and can be completed in less than 15 min, it has a low morbidity and it can be performed on an outpatient basis. The goal of several investigators has been to find alternate implant materials that would be safe for human use [7].

The ideal substance for the endoscopic treatment of reflux and incontinence should be injectable, non-antigenic, non-migratory, volume-stable, and safe for human use. Toward this goal, long-term studies were conducted to determine the effect of injectable chondrocytes in vivo [5]. It was initially determined that alginate, a liquid polysaccharide composed of gluronic and mannuronic acid, embedded with chondrocytes, could serve as a synthetic substrate for the injectable delivery and maintenance of cartilage architecture in vivo. Alginate undergoes hydrolytic biodegradation and its degradation time can be varied depending on the concentration of each of the polysaccharides. The use of autologous cartilage for the treatment of vesicoureteral reflux in humans would satisfy all the requirements for an ideal injectable substance. A biopsy of the ear could be easily and quickly performed, followed by chondrocyte processing and endoscopic injection of the autologous chondrocyte suspension for the treatment of reflux.

Chondrocytes can be readily grown and expanded in culture. Neocartilage formation can be achieved in vitro and in vivo using chondrocytes cultured on synthetic biodegradable polymers [5]. In these experiments, the cartilage matrix replaced the alginate as the polysaccharide polymer underwent biodegradation. This system was adapted for the treatment of vesicoureteral reflux in a porcine model [6].

Six mini-swine underwent surgery to create bilateral reflux. All six were found to have bilateral reflux without evidence of obstruction 3 months after the procedure. Chondrocytes were harvested from the left auricular surface of each mini-swine and expanded with a final concentration of 50×10^6–150×10^6 viable cells per animal. The animals underwent endoscopic repair of reflux with the injectable autologous chondrocyte solution on the right side only. Serial cystograms showed no evidence of reflux on the treated side and persistent reflux in the uncorrected control ureter in all animals. All animals had a successful cure of reflux in the repaired ureter without evidence of hydronephrosis on excretory urography. The harvested ears had evidence of cartilage regrowth within 1 month of chondrocyte retrieval.

At the time of sacrifice, gross examination of the bladder injection site showed a well-defined rubbery-to-hard cartilage structure in the subureteral region. Histologic examination of these specimens showed evidence of normal cartilage formation. The polymer gels were progressively replaced by cartilage with increasing time. Aldehyde fuschin–alcian blue staining suggested the presence of chondroitin sulfate. Microscopic analyses of the tissues surrounding the injection site showed no inflammation. Tissue sections from the bladder, ureters, lymph nodes, kidneys, lungs, liver and spleen showed no evidence of chondrocyte or alginate migration, or granuloma formation. These studies showed that chondrocytes can be easily harvested and combined with alginate in vitro, the suspension can be easily injected cystoscopically and the elastic cartilage tissue formed is able to correct vesicoureteral reflux without any evidence of obstruction [6].

Two multicenter clinical trials were conducted using the above engineered chondrocyte technology. Patients with vesicoureteral reflux were treated at ten centers throughout the US. The patients had a similar success rate as with other injectable substances in terms of cure (Fig. 3). Chondrocyte formation was not noted in patients who had treatment failure. It can safely be supposed that the patients who were cured would have a biocompatible region of engineered autologous tissue present, rather than a foreign material [87]. Patients with

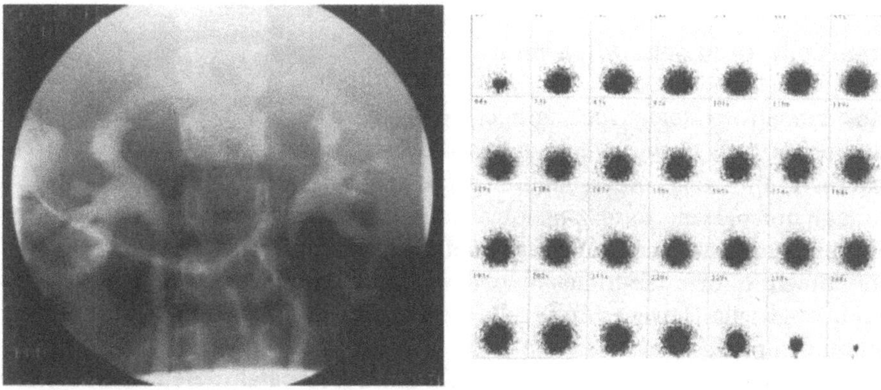

Fig. 3 *Left* Pre-operative voiding cystourethrogram of a patient showing bilateral reflux; *Right* Post-operative radionuclide cystogram of the same patient 6 months after the injection of autologous chondrocytes

urinary incontinence were also treated endoscopically with injected chondrocytes at three different medical centers. Phase 1 trials showed an approximate success rate of 80% at both 3 and 12 months post-operatively [88].

The potential use of injectable, cultured myoblasts for the treatment of stress urinary incontinence has been investigated [89, 90]. Primary myoblasts obtained from mouse skeletal muscle were transfected in vitro to carry the β-galactosidase reporter gene and were then incubated with fluorescent microspheres which would serve as markers for the original cell population. Cells were then directly injected into the proximal urethra and lateral bladder walls of nude mice with a micro-syringe in an open surgical procedure. Tissue was harvested up to 35 days post injection, analyzed histologically and assayed for β-galactosidase expression. Myoblasts expressing β-galactosidase and containing fluorescent microspheres were found at each of the retrieved time points. In addition, regenerative myofibers expressing β-galactosidase were identified within the bladder wall. By 35 days post injection, some of the injected cells expressed the contractile filament α-smooth muscle actin, suggesting the possibility of myoblastic differentiation into smooth muscle. The authors reported that a significant portion of the injected myoblast population persisted in vivo. Similar techniques of sphincter-derived muscle cells have been used for the treatment of urinary incontinence. Strasser harvested muscle samples from pigs, dissociated the cells, and injected autologous pure clones of myoblasts into the urethral wall of pigs under sonographic visualization [91]. Postoperatively, maximal urethral closure pressures were increased markedly in most pigs and the zone of higher urethral closure pressure was lengthened compared to pre-operative measurements. The fact that myoblasts can be transfected, survive after injection and begin the process of myogenic differentiation, further supports the feasibility of using cultured cells of muscular origin as an injectable bioimplant.

9 Stem Cells for Regenerative Medicine

Most current strategies for engineering urologic tissues involve harvesting of autologous cells from the host diseased organ. However, in situations wherein extensive end-stage organ failure is present, a tissue biopsy may not yield enough normal cells for expansion. Under these circumstances, the availability of pluripotent stem cells may be beneficial. Pluripotent embryonic stem cells are known to form teratomas in vivo, which are composed of a variety of differentiated cells. However, these cells are immunocompetent, and would require immunosuppression if used clinically.

The possibility of deriving stem cells from postnatal mesenchymal tissue from the same host, and inducing their differentiation in vitro and in vivo was investigasted. Stem cells were isolated from human foreskin-derived fibroblasts. Stem cell derived chondrocytes were obtained through a chondrogenic lineage process. The cells were grown, expanded, seeded onto biodegradable scaffolds, and implanted in vivo, where they formed mature cartilage structures. This was the first demonstration that stem cells can be derived from postnatal connective tissue and can be used for engineering tissues in vivo ex situ [92].

A second approach which has been pursued for stem lineage isolation involves the isolation of stem cells from individual organs. For example, daily female hormone supplementation is used widely, most commonly in post-menopausal women. A continuous and unlimited hormonal supply produced from ovarian granulosa cells would be an attractive alternative. The feasibility of isolating functional human ovarian granulosa stem cells, which, unlike primary cells, may have the ability to proliferate and function indefinitely, was investigated.

Granulosa stem cells were selectively isolated from post-menopausal human ovaries and their phenotype was confirmed with the stem cell marker antibodies, CD 34, CD 105, and CD [90]. The granulosa stem cells in culture showed steady state progesterone (5–7 ng mL^{-1}) and estradiol (2,500–3,000 pg mL^{-1}) production, either with or without hCG stimulation [93].

10 Therapeutic Cloning

Nuclear transplantation ("therapeutic cloning") could theoretically provide a limitless source of cells for regenerative therapy. According to data from the Centers for Disease Control, as many as 3,000 Americans die every day from diseases that in the future may be treatable with embryonic stem (ES)-derived tissues [94]. In addition to generating functional replacement cells such as cardiomyocytes and neurons, there is also the possibility that these cells could be used to reconstitute more complex tissues and organs, including kidneys [12, 95, 96]. Somatic cell nuclear transfer (SCNT) has the potential to eliminate immune responses associated with the transplantation of these various tissues, and thus the requirement for immunosuppressive drugs and/or immunomodulatory protocols that carry the risk of a wide variety of serious and potentially life-threatening complications (Fig. 4) [97].

Although the goal of "therapeutic" cloning is to generate replacement cells and tissues that are genetically identical with the donor, numerous studies have shown that animals produced by the SCNT technique inherit their mitochondria entirely or in part from the recipient oocyte and not the donor cell [98–100]. This raises the question of whether non-self mitochondrial proteins in cells could lead to immunogenicity after transplantation and defeat the main objective of the procedure.

We tested the histocompatibility of nuclear-transfer-generated cells and engineered tissues in a large animal model, the cow (*Bos taurus*). Cloned muscle cell implants were not rejected, and they remained viable after being transplanted back into the nuclear donor animal, despite expressing a different

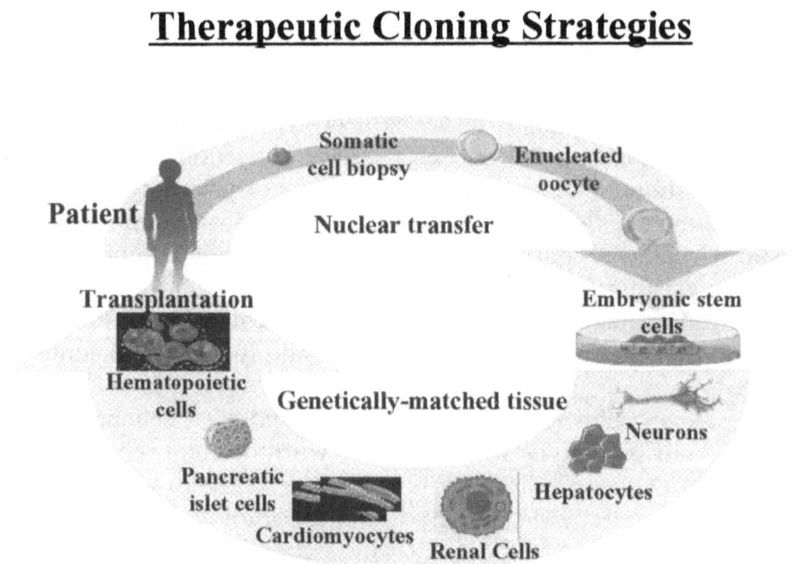

Fig. 4 Therapeutic cloning strategy and its application to the engineering of tissues and organs

mtDNA haplotype. We also showed that nuclear transplantation can be used to generate functional renal structures. Owing to its complex structure and function [17], the kidney is one of the most challenging organs to reconstruct in the body. Previous efforts at tissue-engineering the kidney have been directed toward development of an extracorporeal renal support system comprising both biologic and synthetic components [101–103]. This approach was first described by Aebischer et al. [104, 105], and is being focused towards the treatment of acute rather than chronic renal failure. Humes et al. [101] have shown that the combination of hemofiltration and a renal-assist device containing tubule cells can replace certain physiologic functions of the kidney when they are connected in an extravascular perfusion circuit in uremic dogs. Heat exchangers, flow and pressure monitors, and multiple pumps are required for optimal functioning of this device [106, 107]. Although ex vivo organ substitution therapy would be life-sustaining, there would be obvious benefits for patients if such devices could be implanted long-term without the need for an extracorporeal perfusion circuit or immunosuppressive drugs and/or immune modulatory protocols. While synthetic, selectively permeable barriers can be used ex vivo to separate transplanted cells from the immune system of the body, the implantation of such immunoisolation systems would pose significant difficulties in both the long and short terms [108–111]. We demonstrated that it may be feasible to use therapeutic cloning to generate functional immune-compatible renal tissues [112].

Dermal fibroblasts were isolated from adult Holstein steers by ear notch. Bovine oocytes were obtained from abattoir-derived ovaries. The oocytes were mechanically enucleated at 18–22 h postmaturation, and complete enucleation of the metaphase plate was confirmed with bisBenzimide dye under fluorescence microscopy. A suspension of actively dividing cells was prepared immediately prior to nuclear transfer. Single donor cells were selected and transferred into the perivitelline space of the enucleated oocytes. Fusion of the cell–oocyte complexes was accomplished by applying a single pulse of 2.4 kV cm^{-1} for 15 µs. Nuclear transfer embryos were activated with exposure to Ionomycin. The resulting blastocysts were non-surgically transferred into progestrin-synchronized recipients. The cloned renal and muscle cells were isolated and expanded in vitro after 12 weeks. The expanded cloned renal cells were successfully seeded onto renal units, and implanted back into the nuclear donor organism without immune destruction. The cells organized into glomeruli- and tubule-like structures with the ability to excrete toxic metabolic waste products through a urine-like fluid [112].

11 Muscle

Tissue-engineered constructs containing bovine muscle cells seeded onto PGA matrices were transplanted subcutaneously and retrieved 6 weeks after implantation. After retrieval of the first-set implants, a second set of constructs from the same donor were transplanted for a further 12 weeks. On a histological level, the cloned muscle tissue appeared intact, and showed a well-organized cellular orientation with spindle-shaped nuclei. Immunohistochemical analysis identified muscle fibers within the implanted constructs. In contrast to the cloned implants, the allogeneic, control cell implants failed to form muscle bundles, and showed an increased number of inflammatory cells, fibrosis, and necrotic debris consistent with acute rejection. Semi-quantitative RT-PCR and Western blot analysis confirmed the expression of muscle-specific mRNA and proteins in the retrieved tissues, despite the presence of allogeneic mitochondria. In contrast, expression intensities were significantly lower or absent in constructs generated from genetically unrelated cattle [112].

Immunocytochemical analysis using CD4- and CD8-specific antibodies identified an approximately twofold increase in CD4+ and CD8+ T cells within the explanted first- and second-set control versus cloned constructs. Importantly, the first- and second-set cloned constructs exhibited comparable levels of CD4 and CD8 expression, arguing against the presence of an enhanced second-set reaction as would be expected if mtDNA-encoded minor antigen differences were present. Western blot analysis of the first-set explants indicated an approximately sixfold increase in expression intensity of CD4 in the control versus cloned constructs at 6 weeks, confirming a primary immune response to the control grafts. There was also a significant increase in the mean expression intensities of CD8 in the control versus cloned constructs at 6 weeks. Mean expression intensities of CD4 and CD8 continued to remain significantly elevated in the control versus cloned constructs 12 weeks after second-set implantation [112].

12 Kidney

In the same study, the renal cells obtained through nuclear transfer demonstrated immunochemically the expression of renal specific proteins, including synaptopodin (produced by podocytes), aquaporin 1 (AQP1, produced by proximal tubules and the descending limb of the loop of Henlé), aquaporin 2 (AQP2, produced by collecting ducts), Tamm-Horsfall protein (produced by the ascending limb of the loop of Henlé), and factor VIII (produced by endothelial cells). Synaptopodin- and AQP1- and AQP2-expressing cells exhibited circular and linear patterns in two-dimensional culture respectively. After expansion, the renal cells were shown to produce both erythropoietin and 1,25-dihydroxyvitamin D3, a key endocrinologic metabolite. The cloned cells produced erythropoietin and were responsive to hypoxic stimulation [112].

The cloned renal cells were seeded onto collagen-coated cylindrical polycarbonate membranes. Renal devices with collecting systems were constructed by connecting the ends of three membranes with catheters that terminated in a reservoir (Fig. 5). Thirty-one units (n=19 with cloned cells, n=6 without cells, and n=6 with cells from an allogeneic control fetus) were transplanted subcutaneously and retrieved 12 weeks after implantation back into the nuclear donor animal [112].

On gross examination, the explanted units appeared intact, and straw-yellow colored fluid could be observed in the reservoirs of the cloned group. There was a sixfold increase in volume in the experimental group versus the control groups. Chemical analysis of the fluid suggested unidirectional secretion and concentration of urea nitrogen and creatinine [112].

Physiological function of the implanted units was further evidenced by analysis of the electrolyte levels in the collected fluid as well as specific gravity and glucose concentrations. The electrolyte levels detected in the fluid of the experimental group were significantly different from plasma or the controls. These findings indicate that the implanted renal cells possess filtration, reabsorption and secretory functions. Urine specific gravity is an indicator of kidney function and reflects the action of the tubules and collecting ducts on the glomerular filtrate by furnishing an estimate of the number of particles dissolved in the urine. The urine-specific gravity of cattle is reported as approximately 1.025 (vs 1.027±0.001 for the fluid that was produced by the cloned renal units), and normally ranges from 1.020 to 1.040 (vs approximately 1.010 in normal bovine serum) [34, 35]. The normal range of urine pH for adult herbivores is alkaline, with values ranging from 7.0 to 9.0 [35] (the pH of the fluid from the cloned renal units was 8.1±0.20). Glucose is reabsorbed in the proximal tubules, and is seldom present in the urine of cattle. Glucose was undetectable

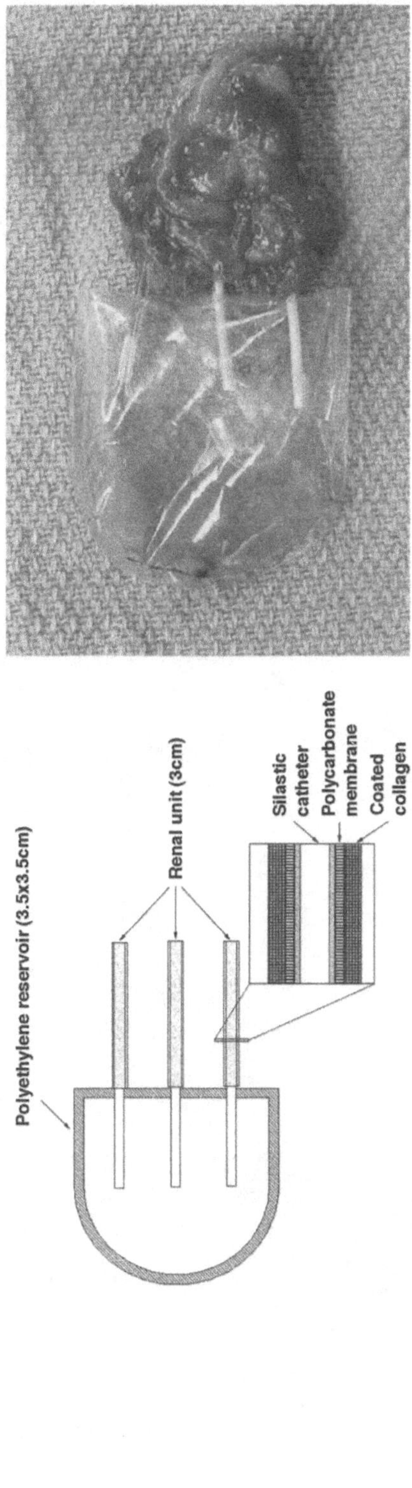

Fig. 5 Tissue-engineered renal units. Illustration of renal unit (*left*) and unit seeded with cloned cells, showing the accumulation of urinelike fluid, retrieved 3 months after implantation (*right*)

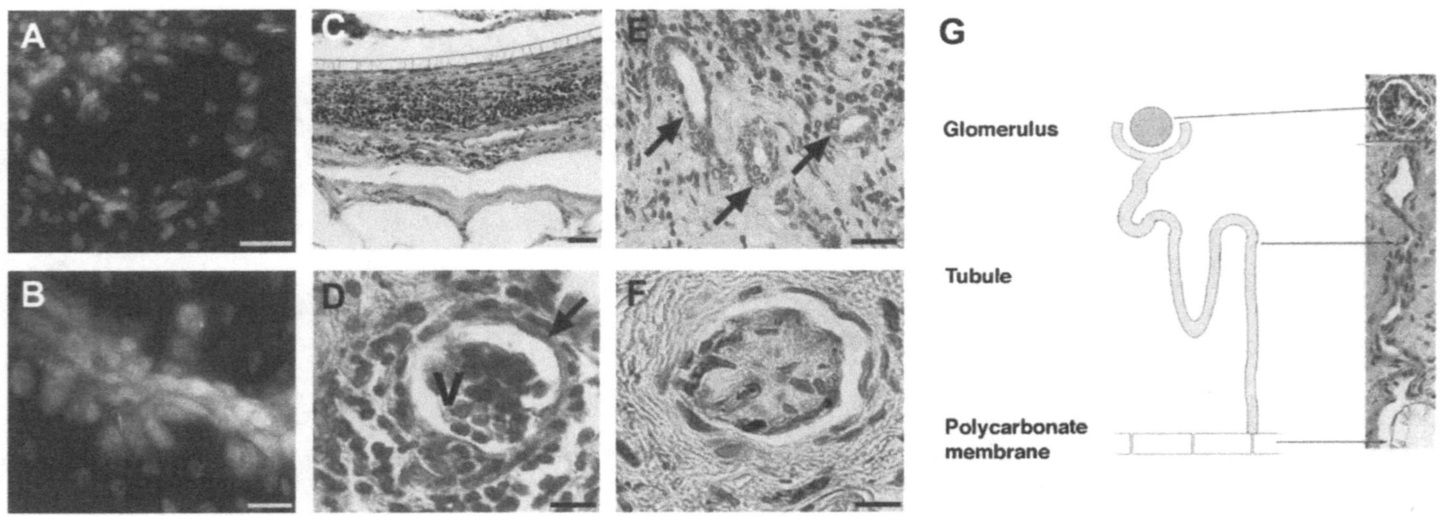

Fig. 6A–G Characterization of renal explants. A Cloned cells stained positively with synaptopodin antibody (A) and AQP1 antibody (B). The allogeneic controls displayed a foreign body reaction with necrosis (C). Cloned explant shows organized glomeruli (D) and tubule (E)-like structures. H&E reduced from 400×. Immunohistochemical analysis using factor VIII antibodies identifies vascular structure within D (F). Reduced from ×400. G There was a clear unidirectional continuity between the mature glomeruli, their tubules, and the polycarbonate membrane

(<10 mg dL^{-1}) in the cloned renal fluid (vs blood glucose concentrations of 76.6±0.04 mg dL^{-1}) [112].

The retrieved implants demonstrated extensive vascularization, and had self-assembled into glomeruli and tubule-like structures. The latter were lined with cuboid epithelial cells with large, spherical and pale-stained nuclei, whereas the glomeruli structures exhibited a variety of cell types with abundant red blood cells. There was a clear continuity between the mature glomeruli, their tubules, and the polycarbonate membrane. The renal tissues were integrally connected in a unidirectional manner to the reservoirs, resulting in the excretion of dilute urine into the collecting systems [112].

Immunohistochemical analysis confirmed expression of renal-specific proteins, including AQP1, AQP2, synaptopodin, and factor VIII. Antibodies for AQP1, AQP2, and synaptopodin identified tubular, collecting tubule, and glomerular segments within the constructs, respectively. In contrast, the allogeneic controls displayed a foreign body reaction with necrosis, consistent with the finding of acute rejection. RT-PCR analysis confirmed the transcription of AQP1, AQP2, synaptopodin, and Tamm-Horsfall genes exclusively in the cloned group. Cultured and cloned cells also expressed high protein levels of AQP1, AQP2, synaptopodin, and Tamm-Horsfall protein, as determined by Western blot analysis. Expression intensity of CD4 and CD8, markers for inflammation and rejection, were also significantly higher in the control versus the cloned group (Fig. 6) [112].

12.1 Mitochondrial DNA Analysis

Previous studies showed that bovine clones harbor the oocyte mtDNA [98–113]. Differences in mtDNA-encoded proteins expressed by clone cells could stimulate a T cell response specific for mtDNA-encoded minor histocompatibility antigens (miHA) [114] when clone cells are transplanted back to the original nuclear donor. The most straightforward approach to resolving the question of miHA involvement is the identification of potential antigens by nucleotide sequencing of the mtDNA genomes of the clone and fibroblast nuclear donor. The contiguous segments of mtDNA that encode 13 mitochondrial proteins and tRNAs were amplified by PCR from total cell DNA in five overlapping segments [112]. These amplicons were directly sequenced on one strand with a panel of sequencing primers spaced at 500 bp intervals. The resulting nucleotide sequences (13,210 bp) revealed nine nucleotide substitutions for the first donor:recipient combination (muscle constructs). One substitution was in the tRNA-Gly segment and five substitutions were synonymous. The sixth substitution, in the *ND1* gene, was heteroplasmic in the nuclear donor where one of the two alternative nucleotides was shared with the clone. A Leu or Arg would be translated at this position in *ND1*. The eighth and ninth substitutions resulted in amino acid (AA) interchanges of Asn>Ser and Val>Ala in the *ATPase6* and *ND4L* genes, respectively. For the second donor:recipient combination (renal constructs), we obtained 12,785 bp from both the clone and nuclear donor animal. The resulting sequences revealed six nucleotide substitutions. One substitution was in the tRNA-Arg segment and three substitutions were synonymous. The fifth and sixth substitutions resulted in AA interchanges of Ile>Thr and Thr>Ile in the *ND2* and *ND5* genes, respectively. The identification of two AA substitutions that distinguish the clone and the nuclear donor confirm that a maximum of only two miHA peptides could be defined by the second donor: recipient combination. Given the lack of knowledge concerning peptide binding motifs for bovine MHC class I molecules, there is no reliable method to predict the impact of these AA substitutions on the ability of mtDNA-encoded peptides to either bind to bovine class I molecules or activate CD8+ cytotoxic T lymphocytes (CTLs) [112].

Although the cloned renal cells derived their nuclear genome from the original fibroblast donor, their mtDNA was derived from the original recipient oocyte. A relatively limited number of mtDNA polymorphisms have been shown to define maternally transmitted miHA in mice115. This class of miHA has been shown to stimulate both skin allograft rejection in vivo and expansion of CTLs in vitro [115], and could constitute a barrier to successful clinical use of such cloned devices as hypothesized for chronic rejection of MHC-matched human renal transplants [116, 117]. We chose to investigate a possible anti-miHA T cell response to the cloned renal devices through both delayed-type hypersensitivity (DTH) testing in vivo and Elispot analysis of IFN-γ-secreting T cells in vitro. An in vivo assay of anti-miHA immunity was chosen based on the ability of skin allograft rejection to detect a wide range of miHA in mice with survival times exceeding 10 weeks [118] and the relative insensitivity of in vitro assays in detecting miHA incompatibility, highlighted by the requirement for in vivo priming to generate CTLs [119]. We were unable to discern an immunological response directed against the cloned cells by DTH testing in vivo. Cloned and control allogeneic cells were intra-dermally injected back into the nuclear donor animal 80 days after the initial transplantation. A positive DTH response was observed after 48 h for the allogeneic control cells but not the cloned cells [112].

The results of DTH analysis were mirrored by Elispot-derived estimates of the frequencies of T cells that secreted IFN-γ following in vitro stimulation. PBLs were harvested from the transplanted recipient 1 month after retrieval of the devices. These PBLs were stimulated in primary mixed lymphocyte cultures (MLCs) with allogeneic renal cells, cloned renal cells, and nuclear donor fibroblasts. Surviving T cells were re-stimulated in anti-IFN-γ-coated wells with either nuclear donor fibroblasts (autologous control) or the respective stimulators used in the primary MLCs. Elispot analysis revealed a relatively strong T cell response to allogeneic renal stimulator cells relative to the responses to either cloned renal cells or nuclear donor fibroblasts. These results corroborate both the relative CD4 and CD8 expression in Western blots as well as the results of in vivo DTH testing to support the conclusion that there was no detectable rejection response that was specific for cloned renal cells following either primary or secondary challenge. Our results suggest that cloned cells and tissues can be grafted back into the nuclear donor organism without immune destruction. These were the first proof-of-principle studies to demonstrate that therapeutic cloning is feasible [112].

13 Conclusion

Regenerative medicine efforts are currently being undertaken for every type of tissue and organ within the urinary system. Most of the effort expended to 'engineer genitourinary tissues has occurred within the last decade. Tissue engineering techniques require a cell culture facility designed for human application. Personnel who have mastered the techniques of cell harvest, culture and expansion as well as polymer design are essential for the successful application of this technology. Various engineered genitourinary tissues are at different stages of development, with some already being used clinically, a few in pre-clinical trials, and some in the discovery stage. Recent progress suggests that engineered urologic tissues may have an expanded clinical applicability in the future.

References

1. Carrel A, Lindbergh C (1938) The culture of organs. Hoeber, New York
2. Atala A (1997) Tissue engineering in the genitourinary system. In: Atala A, Mooney D (eds) Tissue engineering. Birkhauser, Boston, p 149
3. Atala A (1995) J Urol 156:338
4. Atala A (1998) J Urol 159:2
5. Atala A, Cima LG, Kim W, Paige KT, Vacanti JP, Retik AB, Vacanti CA (1993) J Urol 150:745
6. Atala A, Kim W, Paige KT, Vacanti CA, Retik AB (1994) J Urol 152:641
7. Kershen RT, Atala A (1999) Urol Clin 26:81
8. Atala A, Peters CA, Retik AB, Mandell J (1992) J Urol 148:724
9. Atala A, Freeman MR, Vacanti JP, Shepard J, Retik AB (1993) J Urol 150:608
10. Cilento BG, Freeman MR, Schneck FX, Retik AB, Atala A (1994) J Urol 152:655
11. Yoo JJ, Atala A (1997) J Urol 158:1066
12. Machluf M, Atala A (1998) Graft 1:31
13. Yoo JJ, Lee I, Atala A (1998) J Urol 160:1164
14. Fauza DO, Fishman S, Mehegan K, Atala A (1998) J Pediatr Surg 33:377
15. Yoo JJ, Meng J, Oberpenning F, Atala A (1998) Urology 51:221
16. Fauza DO, Fishman S, Mehegan K, Atala A (1998) J Pediatr Surg 33:7
17. Amiel GE, Atala A (1999) Urol Clin 26:235
18. Yoo J, Park H, Lee I, Atala A (1999) J Urol 162:1119
19. Park HJ, Kershen R, Yoo J, Atala A (1999) J Urol 162:1106
20. Oberpenning FO, Meng J, Yoo J, Atala A (1999) Nat Biotechnol 17:2
21. Atala A (1999) Urol Clin 26:157
22. Scriven SD, Booth C, Thomas DF, Trejdosiewicz LK, Southgate J (1997) J Urol 158: 1147

23. Liebert M, Hubbel A, Chung M, Wedemeyer G, Lomax MI, Hegeman A, Yuan TY, Brozovich M, Wheelock MJ, Grossman HB (1997) Differentiation 61:177
24. Puthenveettil JA, Burger MS, Reznikoff CA (1999) Adv Exp Med Biol 462:83
25. Liebert M, Wedemeyer G, Abruzzo LV, Kunkel SL, Hammerberg C, Cooper KD, Grossman HB (1991) Semin Urol 9:124
26. Tobin MS, Freeman MR, Atala A (1994) Surg Forum 45:786
27. Freeman MR, Yoo JJ, Raab G, Soker S, Adam RM, Schneck FX, Renshaw AA, Klagsbrun M, Atala A (1997) J Clin Invest 99:1028
28. Nguyen HT, Park JM, Peters CA, Adam RA, Orsola A, Atala A, Freeman MR (1999) In Vitro Cell Dev Bio 35:371
29. Harriss DR (1995) Br J Urol 75:18
30. Solomon LZ, Jennings AM, Sharpe P, Cooper AJ, Malone PS (1998) J Lab Clin Med 132:279
31. Lobban ED, Smith BA, Hall GD, Harnden P, Roberts P, Selby PJ, Trejdosiewicz LK, Southgate J (1998) Am J Pathol 153:1957
32. Rackley RR, Bandyopadhyay SK, Fazeli-Matin S, Shin MS, Appell R (1999) J Urol 162:1812
33. Bergsma JE, Rozema FR, Bos RRM, van Rozendaal AWM, de Jong WH, Teppema JS, Joziasse CAP (1995) Mater Med 6:715
34. Kim BS, Mooney DJ (1998) Trend Biotechnol 16:224
35. Pariente JL, Kim BS, Atala A (2001) J Biomed Mater Res 55:33
36. Pariente JL, Kim BS, Atala A (2002) J Urol 167:1867
37. Li ST (1995) Biologic biomaterials: tissue-derived biomaterials (collagen). In: Brozino JD (ed) The biomedical engineering handbook. CRC Press, Boca Raton, p 627
38. Silver FH, Pins G (1992) J Long-term Effects Med Implants 2:67
39. Sam AE, Nixon AJ (1995) Osteoarthritis Cartilage 3:47
40. Smidsrød O, Skjåk-Braek G (1990) Trend Biotechnol 8:71
41. Lim F, Sun AM (1980) Science 210:908
42. Piechota HJ, Dahms SE, Nunes LS, Dahiya R, Lue TF, Tanagho EA (1998) J Urol 159:1717
43. Dahms SE, Piechota HJ, Dahiya R, Lue TF, Tanagho EA (1998) Br J Urol 82:411
44. Chen F, Yoo JJ, Atala A (1999) Urology 54:407
45. Gilding DK (1981) Biodegradable polymers. In: Williams DF (ed) Biocompatibility of clinical implant materials. CRC, Boca Raton, p 209
46. Freed LE, Vunjak-Novakovic G, Biron RJ, Eagles DB, Lesnoy DC, Barlow SK, Langer R (1994) Bio/Technology 12:689
47. Mikos AG, Thorsen AJ, Czerwonka LA, Bao Y, Langer R, Winslow DN, Vacanti JP (1994) Polymer 35:1068
48. Harris LD, Kim BS, Mooney DJ (1998) J Biomed Mater Res 42:396
49. Peppas NA, Langer R (1994) Science 263:1715
50. Bazeed MA, Thüroff JW, Schmidt RA, Tanagho EA (1983) Urology 21:53
51. Atala A, Vacanti JP, Peters CA, Mandell J, Retik AB, Freeman MR (1992) J Urol 148:658
52. Olsen L, Bowald S, Busch C et al. (1992) Scand J Urol Nephrol 26:323
53. Kropp BP, Ludlow JK, Spicer D, Rippy MK, Badylak SF, Adams MC, Keating MA, Rink RC, Birhle R, Thor KB (1998) Urology 52:138
54. Sievert KD, Bakircioglu ME, Nunes L, Tu R, Dahiya R, Tanagho EA (2000) J Urol 163:1958
55. ElKassaby AW, Retik AB, Yoo JJ, Atala A (2003) J Urol 169:170
56. DeFilippo RE, Yoo JY, Chen F, Atala A (2002) J Urol 168:1789
57. DeFilippo RE, Pohl HG, Yoo JJ, Atala A (2002) J Urol, 167S:152
58. Neuhof H (1917) Surg Gynecol Obstet 25:383
59. Tsuji I, Ishida H, Fujieda J (1961) J Urol 85:42

60. Kambic H, Kay R, Chen JF, Matsushita M, Harasaki H, Zilber S (1992) J Urol 148:539
61. Kelami A, Ludtke-Handjery A, Korb G, Roll J, Schnell J, Danigel KH (1970) Eur Surg 2:195
62. Fishman IJ, Flores FN, Scott B, Spjut HJ, Morrow B (1987) J Urol 138:1291
63. Probst M, Dahiya R, Carrier S, Tanagho EA (1997) Br J Urol 79:505
64. Sutherland RS, Baskin LS, Hayward SW, Cunha GR (1996) J Urol 156:571
65. Kropp BP, Sawyer BD, Shannon HE, Rippy MK, Badylak SF, Adams MC, Keating MA, Rink RC, Thor KB (1996) J Urol 156:599
66. Vaught JD, Kroop BP, Sawyer BD, Rippy MK, Badylak SF, Shannon HE, Thor KB (1996) J Urol 155:374
67. Probst M, Dahiya R, Carrier S, Tanagho EA (1997) Br J Urol 79:505
68. Kropp BP, Rippy MK, Badylak SF, Adams MC, Keating MA, Rink RC, Thor KB (1996) J Urol 155:2098
69. Lai JY, Yoo JJ, Wulf T, Atala A (2002) J Urol 167S:257
70. Brown AL, Farhat W, Merguerian PA, Wilson GJ, Khoury AE, Woodhouse KA (2002) Biomaterials 23:2179
71. Reddy PP, Barrieras DJ, Wilson G, Bagli DJ, McLorie GA, Khoury AE, Merguerian PA (2000) J Urol 164:936
72. Merguerian PA, Reddy PP, Barrieras DJ, Wilson GJ, Woodhouse K, Bagli DJ, McLorie GA, Khoury AE (2000) BJU Int 85:894
73. Portis AJ, Elbahnasy AM, Shalhav AL, Brewer A, Humphrey P, McDougall EM, Clayman RV (2000) J Urol 164:1405
74. Portis AJ, Elbahnasy AM, Shalhav AL, Brewer AV, Olweny E, Humphrey PA, McDougall EM, Clayman RV (2000) J Endourol 14:203
75. Gleeson MJ, Griffith DP (1992) J Urol 148:1377
76. Bona AV, De Gresti A (1966) Minerva Urol 18:43
77. Monsour MJ, Mohammed R, Gorham SD, French DA, Scott R (1987) Urology 15:235
78. Rohrmann D, Albrecht D, Hannappel J, Gerlach R, Schwarzkopp G, Lutzeyer W (1996) J Urol 156:2094
79. De Boer WI, Schuller AG, Vermay M, van der Kwast TH (1994) Am J Pathol 145:1199
80. Baker R, Kelly T, Tehan T, Putman C, Beaugard E (1955) J Am Med Assoc 168:1178
81. Gorham SD, French DA, Shivas AA, Scott R (1989) Eur Urol 16:440
82. Kershen RT, Yoo JJ, Moreland RB, Krane RJ, Atala A (1998) J Urol 159:156
83. Falke G, Yoo J, Machado M, Moreland R, Atala A (in press) J Tissue Eng
84. Kwon TG, Yoo JJ, Atala A (2002) J Urol 168:1754
85. Nukui F, Okamoto S, Nagata M, Kurokawa J Fukui J (1997) Int J Urol 4:52
86. Thomalla JV, Thompson ST, Rowland RG, Mulcahy JJ (1987) J Urol 138:65
87. Diamond DA, Caldamone AA (1999) J Urol 162:1185
88. Bent A, Tutrone R, McLennan M, Lloyd K, Kennelly M, Badlani G (2001) Neurourol Urodynam 20:157
89. Yokoyama T, Chancellor MB, Watanabe T, Ozawa H, Yoshimura N, de Groat WC, Qu, Z, Huard J (1999) J Urol 161:307
90. Chancellor MB, Yokoyama T, Tirney S, Mattes CE, Ozawa H, Yoshimura N, de Groat WC, Huard J (2000) Neurourol Urodynam 19:279
91. Strasser H, Marksteiner R, Eva M, Stanislav B, Guenther K, Helga F et al. (2003) Transurethral ultrasound - experimental results. In: Proceedings of the 2003 International Bladder Symposium, Arlington, VA
92. Bartsch G, Yoo J, Kim B, Atala A (2000) J Urol 1009S:227
93. Raya-Rivera A, Yoo J, Atala A (2000) Hormone producing - intersex disorders. In: American Academy of Pediatrics Meeting, Chicago

94. Lanza RP et al. (2001) Science 293:1299
95. Atala A, Lanza RP (eds) (2001) Methods of tissue engineering. Academic, San Diego
96. Atala A, Mooney D (eds) (1997) Synthetic biodegradable polymer scaffolds. Birkhäuser, Boston
97. Lanza RP, Cibelli JB, West MD (1999) Nat Biotechnol 17:1171
98. Evans MJ et al. (1999) Nat Genet 23:90
99. Hiendleder S, Schmutz SM, Erhardt G, Green RD, Plante Y (1999) Mol Reprod Dev 54:24
100. Steinborn R et al. (2000) Nat Genet 25:255
101. Humes HD, Buffington DA, MacKay SM, Funke AJ, Weitzel WF (1999) Nat Biotechnol 17:451
102. Cieslinski DA, Humes HD (1994) Biotechnol Bioeng 43:781
103. MacKay SM, Kunke AJ, Buffington DA, Humes HD (1998) ASAIO J 44:179
104. Aebischer P, Ip TK, Panol G, Galletti PM (1987) Life Support Syst 5:159
105. Ip T, Aebischer P, Galletti PM (1988) ASAIO Trans 34:351
106. Humes HD (2000) Renal replacement devices. In: Lanza RP, Langer R, Vacanti J (eds) Principles of tissue engineering, 2nd edn. Academic, San Diego, p 645
107. Amiel A, Yoo J, Atala A (2000) World J Urol 18:71
108. Lanza RP, Hayes JL, Chick WL (1996) Nat Biotechnol 14:1107
109. Kuhtreiber WM, Lanza RP, Chick WL (eds) (1999) Cell encapsulation technology and therapeutics. Birkhauser, Boston
110. Lanza RP, Chick WL (eds) (1994) Immunoisolation of pancreatic islets. Landes, Austin
111. Joki T, Machluf M, Atala A, Zhu J, Seyfried NT, Dunn IF, Abe T, Carroll RS, Black PM (2001) Nature Bio 19:35
112. Lanza RP, Chung HY, Yoo JJ, Wettstein P, Blackwell C, Atala A et al. (2002) Nat Biotechnol 20:689
113. Lanza RP et al. (2000) Cloning 2:79
114. Fischer Lindahl K, Hermel E, Loveland BE, Wang CR (1991) Annu Rev Immunol 9:351
115. Fischer Lindahl K, Hermel E, Loveland BE, Wang CR (1991) Annu Rev Immunol 9:351
116. Hadley GA, Linders B, Mohanakumar T (1992) Transplantation 54:537
117. Yard BA et al. (1993) Kidney Int 43:S133
118. Bailey DW (1975) Immunogenetics 2:249
119. Mohanakumar T (1994) The role of MHC and non-MHC antigens in allograft immunity. Landes, Austin, p 1

Received: June 2004

Author Index Volumes 51 – 94

Author Index Volumes 1 – 50 see Volume 50

Subject Index

FSC
www.fsc.org
MIX
Papier | Fördert gute Waldnutzung
FSC® C083411

Kolibri 360 GmbH
Ferdinand-Jühlke-Straße 7
99095 Erfurt, Deutschland
produktsicherheit@kolibri360.de